# Innovation in Global Green Technologies 2020

*Edited by Albert Sabban*

Published in London, United Kingdom

IntechOpen

*Supporting open minds since 2005*

Innovation in Global Green Technologies 2020
http://dx.doi.org/10.5772/intechopen.83271
Edited by Albert Sabban

Contributors
Woodrow Clark, Danilo Bonato, Liliana Naval, Fábio Ribeiro, Yeda Silva, Pham Van Kien, Le Anh Duc, Nguyen Hay, Markandeya Tiwari, Devendra Mohan, S.P. Shukla, Xiaodong Lai, Samuel Adams, William Asante, Boguslaw Bieda, Dariusz Sala, Hadi Guna, Mohammad Syuhaimi Ab-Rahman, Norhana Arsad, Roslan Shukor, Sahbudin Shaari, Albert Sabban



First published in London, United Kingdom, 2020 by IntechOpen
IntechOpen is the global imprint of INTECHOPEN LIMITED, registered in England and Wales, registration number: 11086078, 7th floor, 10 Lower Thames Street, London, EC3R 6AF, United Kingdom
Printed in Croatia

British Library Cataloguing-in-Publication Data
A catalogue record for this book is available from the British Library

Additional hard and PDF copies can be obtained from orders@intechopen.com

Innovation in Global Green Technologies 2020
Edited by Albert Sabban
p. cm.
Print ISBN 978-1-83880-049-9
Online ISBN 978-1-83880-050-5
eBook (PDF) ISBN 978-1-78984-876-2

For EU product safety concerns:
IN TECH d.o.o., Prolaz Marije Krucifikse Kozulić 3, 51000 Rijeka, Croatia, info@intechopen.com or visit our website at intechopen.com.

# Meet the editor

Dr. Albert Sabban received a BSc and MSc Magna Cum Laude in Electrical and Computer Engineering from Tel Aviv University, Israel, in 1976 and 1986, respectively. He received an MBA from Haifa University, Israel. He received a PhD in Electrical and Computer Engineering from Colorado University at Boulder, USA, in 1991. Dr. Sabban's reasearch interests include medical wearable and Internet of Things (IOT) systems, microwave engineering, green technologies, and communication and system engineering. In 1976 he joined Rafael Advanced Defense Systems Ltd. (RAFAEL) in Israel where he worked as a senior researcher, group leader, and project leader until 2007. From 2008 to 2010 he worked as a radio frequencey (RF) specialist and project leader at high-tech companies. Since 2004 he has served as a senior lecturer and researcher at colleges and universities across Israel. He has published more than 100 research papers and holds patents in the field of antennas. Dr. Sabban has written five books on wearable systems and wearable wideband antennas. He also wrote a book on electrmagnetics and microwave theory for graduate students as well as two book chapters on wearable microstrip antennas. Dr. Sabban serves as an editor of several books on wearable systems, green technologies, and antennas.

# Contents

# Preface

The continuous growth in development of cellular wireless communication systems over the last thirty years has resulted in most of the world's population owning smartphones, smart watches, iPads, and other devices. As a result, the number of unwanted electronic devices is growing at a rapid rate. With a huge number of devices being produced and discarded, a new environmental disaster has stricken our planet. Electronic waste or discarded electronics are filling up landfills at an alarming rate. These electronic devices contain hazardous and toxic materials that endanger the environment and the health of local communities. As a result, green technologies for recycling this waste are being employed to decrease environmental pollution. Renewable energy is a major factor in decreasing environmental pollution.

Divided into the following three sections, this book presents comprehensive information on green technologies.

1. Electronic Green Technologies

2. Recycling and Waste Management

3. Global Green Technologies – Economics and Innovation

Section 1 includes chapters on energy harvesting, the green technology of wavelength division multiplexing based on polymer optical fiber (WDM-POF), and radio-frequency generators used in drying technology.

Section 2 includes chapters on auxiliary strategies for water management, minimization of water usage and recycling and reuse of effluent, and management of coal fly ash in the remediation process.

Section 3 includes chapters on strategies in global business economics, green technology innovations in China, and politics of renewable energy in Africa as well as the nature, prospects, and challenges of renewable energy in the continent.

I want to thank all the contributing authors for their excellent work in writing the book's chapters.

**Dr. Albert Sabban**
Kinneret Academic College,
Israel

Ort Braude College,
Karmiel, Israel

Section 1

# Introduction

Chapter 1

# Introductory Chapter: Introduction to Global Green Technologies

*Albert Sabban*

## 1. Introduction

It is obvious in 2020 to most of the world population that green technologies are the salvation army of mother earth and the universe. In the last 50 years, the universe has suffered rapid changes in climate. Our planet suffers from increasingly severe droughts, rising seawater levels, seawater acidification, increased depletion of groundwater reserves, and global rise of earth temperature. The rapid spread of diseases and macro-parasites and the extinction of species are the direct results of climate changes. These changes may be irreversible if countries, governments, and citizens do not act rapidly to save the planet.

The high-tech industrial revolution in the last 50 years depleted and ruined the planet natural resources. Electronic waste, plastics, and food garbage contain dangerous chemicals that pollute soil and groundwater with toxic chemicals that cannot be removed from our drinking water supply. These toxic materials can be found in food crops and livestock grown on contaminated soil. These toxic wastes and air pollution affect significantly our health. The planet oceans, seas, and lakes suffer from plastic pollution. Plastic wastes destroy the ocean habitats of sea creatures around the world. Plastic pollution kills fishes, birds, and other species. Plastic large pieces pose choking and strangulation hazards. Tiny plastic particles are making their way into the bottom of the food chain. Fishes swallow plastic waste and become contaminated. These fishes are subsequently harvested for human consumption. The contaminated fishes may be found in our plate and in our stomach.

Renewable green energy, waste management, and recycling are the main challenges and topics in the research and innovation in green technologies. The book is divided into three sections. Section 1 presents innovations in green electronic technologies. Section 2 presents recycling and waste management. Innovation and economics in global green technologies are presented in Section 3.

## 2. Electronic green technologies

The electronics and electrical industries are facing increasing pressure from legislation to remove.

toxic and hazardous materials from their products. There is a continual worldwide environmental movement to use green materials and green electronics in the production of electronic devices. There are now a series of initiatives worldwide that outline targets for electronic equipment reuse and recycling. In green electronics, the use of hazardous materials such as lead, plastic materials, and other toxic materials is not allowed or limited in order to improve the ease of recycling.


IntechOpen

Electronics manufacturers are faced with the unique challenge to develop electronic devices that are environmentally friendly. Green electronics manufacturing focuses on efficient green energy, usage of green materials, reduction of devices' parts, reusage of green materials, and recycling energy and devices' components. The book will present methods that achieve the production of green efficient energy in electronics industry. In this regard, energy harvesting technologies that can be employed to produce, recycle, and store green energy and green materials that are important factors in developing greener electronics will be discussed. Computers, cell phones, and other electronic devices contain toxic hazardous materials that endanger the environment and, consequently, threaten the human health. The book will present several green materials that are environmentally friendly, and it will try to show how system engineering tools can be applied to develop optimal green electronic systems. Another topic that this book will consider concerns the electronic waste. Electronics manufacturers are encouraged to reduce the amount of electronic waste by designing products that minimize the amount of harmful substances and to use components that can be recycled and also to develop products with longer lifecycle. In conclusion, this book intends to provide the reader with a comprehensive overview of green electronics technologies as a desirable standard for manufacturers.

## 3. Renewable energy

Renewable energy converts energy from natural resources, light and wind, into electrical energy. Examples of renewable energy are solar energy, wind energy, water energy, and biology fuel. Nonrenewable energy resources include nuclear, hydrogen, coal, natural gas, and oil.

### 3.1 Solar energy

Solar cell directly converts energy from natural light into electrical energy via the process of photovoltaics. Generating electricity from solar energy results in less consumption of fossil fuels, as well as the reduction of pollution. **Figure 1** presents a solar energy site.

Advantages of solar energy:

- Green and clean energy.
- Environment-friendly.
- Not degradable.
- Solar light is a free natural resource.

Disadvantages of solar energy:

- Depend on the weather and sunlight.
- Expensive.
- Consume large area.
- Development and production of solar site is expensive.

**Figure 1.**
*Solar energy site in Israel.*

### 3.2 Wind energy

The wind kinetic energy is used to operate electric turbines and windmills. However, windmills cannot be operated in a residential area. Offshore wind turbines provide steady, reliable clean energy in several countries. **Figure 2** presents a wind energy site.

Advantages of wind energy:

- Green and clean energy.
- Not degradable.
- Wind is a free natural resource.
- Cheap energy.

Disadvantages of wind energy:

- Depend on the weather and wind.
- Cannot be operated in a residential area.
- Consume large area.
- Development and production of wind energy site is expensive; land is expensive.

**Figure 2.**
*Wind energy site.*

- Noisy.
- Dangerous to birds and animals.

### 3.3 Water energy

Waterfalls and water flow may be used to generate electric energy. Water energy site is presented in **Figure 3**. Water flow kinetic energy is converted to electric energy.

Sources of water energy:

- Water flow due to rainfall.
- Waterfalls.
- Water stream, high water levels, and low water levels.

Advantages of water energy:

- Green and clean energy.
- Not degradable.
- Water is a free natural resource.
- Cheap energy.

Disadvantages of wind energy:

- Depend on the weather and water stream.
- Cannot be operated in a residential area.
- Limited to locations of waterfalls.
- Consume large area.
- Development and production of water energy site is expensive.

**Figure 3.**
*Water energy site.*

### 3.4 Energy harvesting

In the last decade, the idea of employing free space energy in the forms of heat, light, vibration, electromagnetic waves, muscle motion, and other types of energy has become useful and attractive. A number of methods to produce electricity from these different types of energy sources have been developed [1–3]. Energy harvesting systems may eliminate the need to replace batteries every day and the usage of power cords. In order to use as much free space energy as possible, it is important to collect the electromagnetic power from several wireless communication systems. In these cases, we should use wideband or multiband antennas. The energy harvesting antenna must satisfy several specific requirements related to the system application. Due to considerably low-power densities, highly efficient radiators are crucial. The antennas should operate at a specific frequency range and polarization. The antenna radiation pattern should have a wide beam width or omnidirectional radiation pattern. Several printed antennas were employed for harvesting energy applications [4–10]. In RF energy harvesting systems, electromagnetic waves propagating in free space are captured, stored, and used to charge batteries and for other applications. There is a significant increase in the amount of electromagnetic energy in the air. The expected amount of radio wave in the air in 2013 was 1.5 exabytes per month. However, the expected amount of radio wave in the air in 2017 was 11 exabytes per month. Energy sources used in harvesting systems are listed in **Table 1**.

Harvesting energy concept is presented in **Figure 4**. The RF energy harvesting system consists of an antenna, a rectifying circuit, and a rechargeable battery. The

| Energy Source | Type | Efficiency | Estimated Harvested Power |
|---|---|---|---|
| Light | Outdoor / Indoor | 10~25% | 100 mW/cm$^2$ |
| Thermal | Human<br>Industrial | ~0.1%<br>~3% | 60 μW/cm$^2$<br>~1-10 mW/cm$^2$ |
| Vibration | ~Hz–human<br>~kHz–machines | 20~50% | ~4 μW/cm$^3$<br>~800 μW/cm$^3$ |
| Electromagnetic | 900 - 2700MHz<br>WiFi, WLAN | ~50% | 0.1 μW/cm$^2$<br>0.001 μW/cm$^2$ |

**Table 1.**
*Energy sources used in harvesting systems.*

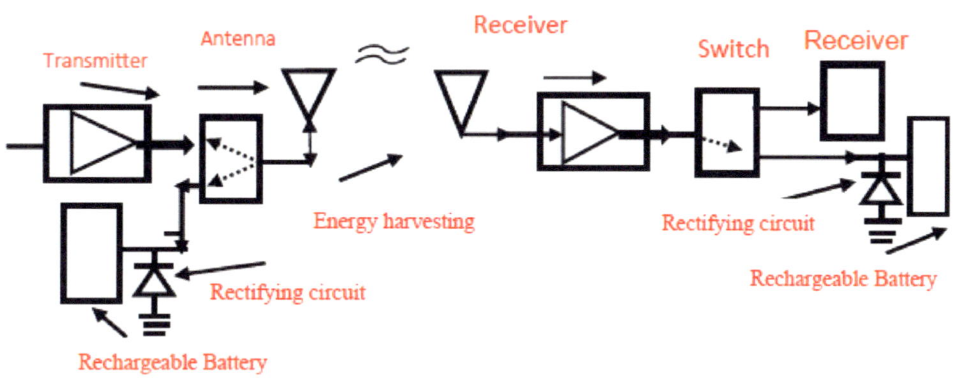


**Figure 4.**
*Dual-mode energy harvesting concept.*

harvesting energy system operates as a dual-mode energy harvesting system. The low-noise amplifier is part of the receiving system. The LNA DC bias voltages are supplied by the receiving system.

## 4. Recycling

Recycling is the process of collecting and processing materials that would otherwise be thrown away as trash and turning them into new products. The continuous growth in the development of cellular wireless communication systems over the last 30 years has resulted in the majority of world population owning smartphones, smart watches, I-pads, and other devices. As a result, the number of unwanted electronic devices is growing at a rapid rate. With this huge number of devices being produced and discarded, a new environmental disaster strikes our planet. Electronic waste or discarded old electronics are filling up landfills at an alarming rate. These electric devices contain hazardous materials and toxic materials that endanger the environment and the health of local communities. This electronic waste increases environmental pollution. Green technologies are employed to decrease environmental pollution. Recycling of electronic waste, old batteries, plastic waste, and bottles decreases environmental pollution. Recycling saves original raw materials. Recycling reduces waste and pollution. Recycling process is presented in **Figure 5**. Recycling outdoor site is presented in **Figure 6**.

Materials recycling:

- Paper recycling.
- Plastic recycling.
- Metal recycling.

**Recycling advantages**:

- Conserves natural resources such as timber, water, and minerals.
- Reduces the amount of waste sent to landfills and incinerators.

**Figure 5.**
*Recycling process.*

**Figure 6.**
*Recycling outdoor site.*

- Increases economic security by tapping a domestic source of materials.
- Prevents pollution by reducing the need to collect new raw materials.
- Saves energy.
- Supports manufacturing and conserves valuable resources.

## 5. Innovations and challenges in green technologies

Electronics developer and manufacturers are now faced with a great challenge of designing, developing, producing, and installing electronics with great attention to the product's environmental impact throughout its entire lifecycle. Electronic devices create a lot of waste. Electronics manufacturers are encouraged to reduce the amount of electronic waste by designing products that minimize the amount of toxic materials used. Electronic companies should recycle and reuse as much as possible raw materials and parts. Green electronic production includes the manufacturing of products with green materials and longer lifecycles. This helps to reduce the amount of electronic waste created by the electronic companies all over the world. Green devices focus on reducing the number of parts, green materials, and recycling components and materials. Green production allows electronics manufacturers to create environmentally friendly products with a long lifecycle with a good quality and reliability.

Future green technologies challenges and innovations

- Producing cheap solar green energy.
- Recycling most of the electronic and material wastes.
- Development and production of green cars and airplanes.

## Author details

Albert Sabban[1,2]

1 Kinneret Academic College, Israel

2 Ort Braude College, Karmiel, Israel

*Address all correspondence to: sabban@mx.kinneret.ac.il

## References

[1] Paradiso JA, Starner T. Energy scavenging for mobile and wireless electronics. IEEE Pervasive Computing. 2005;**4**(1):18-27

[2] Sabban A. Compact wearable meta materials antennas for energy harvesting systems, medical and IOT systems. MDPI Electronics. November 2019

[3] Valenta CR, Durgin GD. Harvesting wireless power: Survey of energy-harvester conversion efficiency in far-field, wireless power transfer systems. IEEE Microwave Magazine. 2014;**15**(4):108-120

[4] Nintanavongsa P, Muncuk U, Lewis DR, Chowdhury KR. Design optimization and implementation for RF energy harvesting circuits. IEEE Journal on Emerging and Selected Topics in Circuits and Systems. 2012;**2**(1):24-33

[5] KKA D, Sadasivam S, Din NM, Chakrabarthy CK. Design of a 377 Ω patch antenna for ambient RF energy harvesting at downlink frequency of GSM 900. In: Proceedings of the 17th Asia Pacific Conference on Communications (APCC '11). Sabah, Malaysia; 2011. pp. 492-495

[6] Rahim RA, Malek F, SFW A, SLS H, Junita MN, Hassan HF. A harmonic suppression circularly polarized patch antenna for an RF ambient energy harvesting system. In: Proceedings of the IEEE Conference on Clean Energy and Technology (CEAT '13). Lankgkawi, Malaysia: IEEE; 2013. pp. 33-37

[7] Krakauskas M, Sabaawi AMA, Tsimenidis CC. Suspended patch microstrip antenna with cut rectangular slots for RF energy harvesting. In: Proceedings of the 10th Loughborough Antennas and Propagation Conference (LAPC '14). Loughborough, UK; 2014. pp. 304-307

[8] Sabban A. Low Visibility Antennas for Communication Systems. USA: Taylor & Francis Group; 2015

[9] Sabban A. Wideband RF Technologies and Antenna in Microwave Frequencies. USA: Wiley Sons; 2016

[10] Sabban A. New wideband passive and active wearable antennas for energy harvesting applications. Journal of Sensor Technology. December 2017. pp. 53-70

Section 2

# Electronic Green Technologies

Chapter 2

# Wideband Passive and Active Wearable Energy Harvesting Systems for Medical and IOT Applications

*Albert Sabban*

## Abstract

Demand for green energy is in continuous growth in the last years. Compact efficient antennas are crucial for energy harvesting portable systems. Small antennas have low efficiency. The efficiency of communication and energy harvesting systems may be improved by using efficient passive and active antennas. The system dynamic range may be improved by connecting amplifiers to the small antenna feed line. Novel passive and active portable harvesting systems are presented in this chapter. Printed patch, notch and Slot antennas are compact and have low volume. The active antennas may be employed in energy harvesting wearable systems. The antennas and the harvesting system components may be assembled on the same printed board. The printed notch and slot antennas bandwidth are from 40 to 100% for VSWR better than 3:1. The slot antenna gain is around 3 dBi with efficiency higher than 85%. The antennas' electrical parameters were computed in free space and near the human body. The active notch antenna gain is around 23 $\pm$ 3dB for frequencies ranging from 200 to 900 MHz. The active notch antenna gain is 13 $\pm$ 3 dB for frequencies ranging from 1 to 3 GHz. The active notch and slot antenna noise figure is 0.5 $\pm$ 0.3dB for frequencies ranging from 200 MHz to 3 GHz.

**Keywords:** energy harvesting systems, wearable sensors, active systems, medical applications, sensor chargers

## 1. Introduction

In the last decade, the idea of employing free space energy in the forms of heat, light, vibration, electromagnetic waves, muscle motion, and other types of energy has become useful and attractive. A number of methods to produce electricity from these different types of energy sources have been developed [1–3]. Energy harvesting systems may eliminate the need to replace batteries everyday and the usage of power cords. In order to use as much free space energy as possible, it is important to collect the electromagnetic power from several wireless communication systems. In these cases, we should use wideband or multiband antennas. The energy harvesting antenna must satisfy several specific requirements related to the system application. Due to considerably low-power densities, highly efficient radiators are crucial. The antennas should operate at a specific frequency range and polarization. The antenna radiation pattern should have a wide beam width or omnidirectional radiation pattern. Several printed antennas were employed for harvesting energy applications [4–6]. Patch and slot antennas are widely used in communication and medical system [7–24]. Wideband compact slot and notch antennas are good choice to function in wearable harvesting energy systems. Slot and notch antennas are compact and flexible and have low production cost. Moreover, a compact low-cost feed network may be achieved by integrating the amplifiers with the antennas on the same substrate. Printed wearable antennas are widely presented in the literature in the last decade as referred in [7–26]. The human body effect on the electrical performance of wearable antennas at microwave frequencies is not always presented in the literature. Electrical properties of human tissues have been investigated in several papers such as [27, 28]. Several wearable antennas have been presented in papers in the last decade as referred in [22, 28–36]. Wearable printed notch and slot antennas for harvesting energy applications are rarely presented in the literature. A new class of wideband passive and active wearable antennas for harvesting energy applications is presented in this chapter. The system efficiency and dynamic range may be improved by connecting amplifiers to the antenna feed line. The active antenna gain is around 23 dB, and the active antenna noise figure is 0.3 dB for frequencies from 200 to 600 MHz.

## 2. Energy harvesting systems

In RF energy harvesting systems, electromagnetic waves propagating in free space are captured, stored, and used to charge batteries and for other applications. There is a significant increase in the amount of electromagnetic energy in the air. The expected amount of radio wave in the air in 2013 was 1.5 exabytes per month. However, the expected amount of radio wave in the air in 2017 was 11 exabytes per month (see **Table 1**). Today we can do more computations per kWh as listed in **Table 2**. Energy sources used in harvesting systems are listed in **Table 3**. Wireless communication systems operate in the frequencies from 700 to 2700 MHz. Medical systems operate in the frequencies from 200 to 1200 MHz. WLAN systems operate in the frequencies from 5400 to 5900 MHz.

RF energy is inversely proportional to distance and therefore drops as the distance from a source is increased. Harvested power from RF energy sources is around 0.1 $\mu W/cm^2$. Harvested power from RF energy sources in malls and stadiums may increase to around 1 $mW/cm^2$. RF energy harvesting concept is shown in **Figure 1**.

The RF energy harvesting system consists of an antenna, a rectifying circuit, and a rechargeable battery. The harvesting energy system operates as a dual mode energy harvesting system. The low-noise amplifier is part of the receiving system. The LNA DC bias voltages are supplied by the receiving system. We can calculate the energy harvesting link budget by using Eqs. (1–4):

| Year | Amount of radio wave in free space exabytes per month |
|---|---|
| 2014 | 2.6 |
| 2015 | 4.4 |
| 2016 | 7 |
| 2017 | 11 |

**Table 1.**
*Amount of radio wave in free space.*

| Year | Computations per kWh (1E+09) |
|---|---|
| 1983 | 10 |
| 1985 | 50 |
| 1987 | 100 |
| 1992 | 1000 |
| 1997 | 10,000 |
| 2003 | 100,000 |
| 2008 | 1,000,000 |
| 2010 | 15,000,000 |

**Table 2.**
*Computations per kWh.*

| Energy source | Type | Efficiency | Estimated Harvested Power |
|---|---|---|---|
| Light | Outdoor/indoor | 10–25% | 100 mW/cm$^2$ |
| Thermal | Human<br>Industrial | ~0.1%<br>~3% | 60 μW/cm$^2$<br>~1–10 mW/cm$^2$ |
| Vibration | ~Hz—human<br>~kHz—machines | 20–50% | ~4 μW/cm$^3$<br>~800 μW/cm$^3$ |
| Electromagnetic | 900–2700 MHz<br>Wi-Fi, WLAN | ~50% | 0.1 μW/cm$^2$<br>0.001 μW/cm$^2$ |

**Table 3.**
*Energy sources used in harvesting systems.*

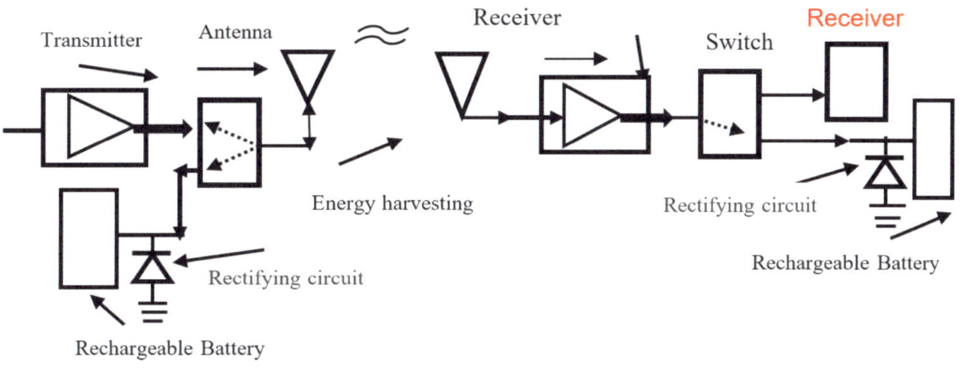


**Figure 1.**
*Dual mode energy harvesting concept.*

$$P_r = P_t G_t G_r \left(\frac{\lambda}{4\pi R}\right)^2 \tag{1}$$

Free-space loss (Lp) represents propagation loss in free space. Losses due to attenuation in atmosphere, La, should also be accounted for in the transmission equation, where, $L_p = \left(\frac{4\pi R}{\lambda}\right)^2 \cdot$ The received power may be given as $P_r = \frac{P_t G_t G_r}{L_p}$.

Losses due to polarization mismatch, $L_{pol}$, should also be accounted. Losses associate with receiving antenna, $L_{ra}$, and with the receiver, $L_r$, cannot be neglected in computation of transmission budget. Losses associate with the transmitting antenna as written as $L_{ta}$.

$$P_r = \frac{P_t G_t G_r}{L_p L_a L_{ta} L_{ra} L_{pol} L_o L_r} \tag{2}$$

$P_t = P_{out}/L_t$, EIRP = $P_t G_t$.

where $P_t$ = transmitting antenna power; $L_t$ = loss between the power source and antenna; EIRP = effective isotropic radiated power.

$$\begin{aligned} P_r &= \frac{P_t G_t G_r}{L_p L_a L_{ta} L_{ra} L_{pol} L_{other} L_r} \\ &= \frac{EIRP \times G_r}{L_p L_a L_{ta} L_{ra} L_{pol} L_{other} L_r} \\ &= \frac{P_{out} G_t G_r}{L_t L_p L_a L_{ta} L_{ra} L_{pol} L_{other} L_r} \end{aligned} \tag{3}$$

where $G = 10 \cdot \log\left(\frac{P_{out}}{P_{in}}\right)$ dB gain in dB; $L = 10 \cdot \log\left(\frac{P_{in}}{P_{out}}\right)$ dB loss in dB.

The received power $P_r$ in dBm is given in Eq. (4). The received power $P_r$ is commonly referred to as "Carrier Power."

$$P_r = EIRP - L_{ta} - L_p - L_a - L_{pol} - L_{ra} - L_{other} + G_r - L_r \tag{4}$$

Wireless smart phone using standard 802.11 can transmit up to 1 W.

PCMCIA cards using standard 802.11 can transmit around 10 mW up to 100 mW.

## 3. Wideband notch antenna, 2–7.8 GHz, for energy harvesting applications

A compact notch antenna was printed on a dielectric substrate with dielectric constant of 2.2. The antenna dimensions are 116.4 × 71.4 × 1.2 mm, as presented in **Figure 2**. The antenna bandwidth for VSWR better than 2.5:1 is around 90–100% (see **Figure 3**). The notch antenna has VSWR better than 3:1 at frequencies from 2.1

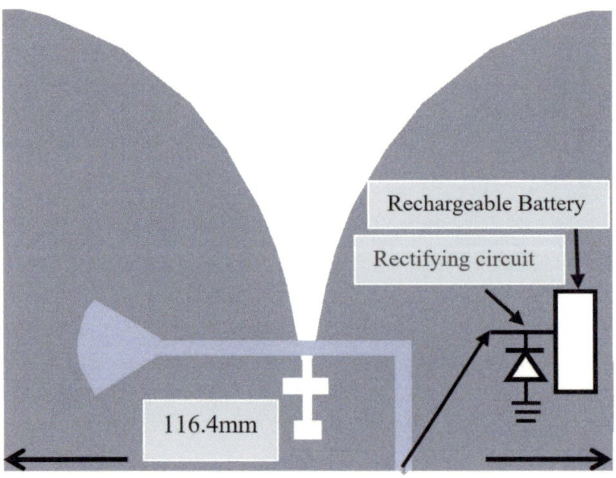


**Figure 2.**
*A wideband 2–7.8 GHz energy harvesting notch.*

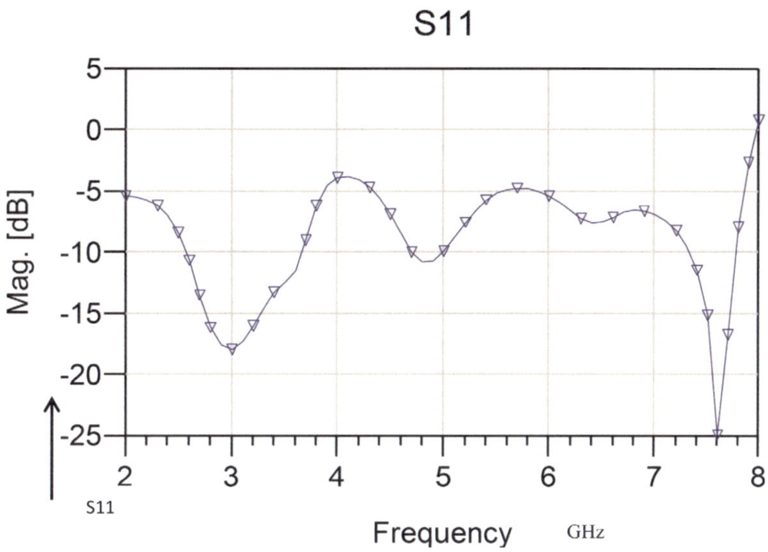


**Figure 3.**
*A wideband 2–7.8 GHz notch, computed S11.*

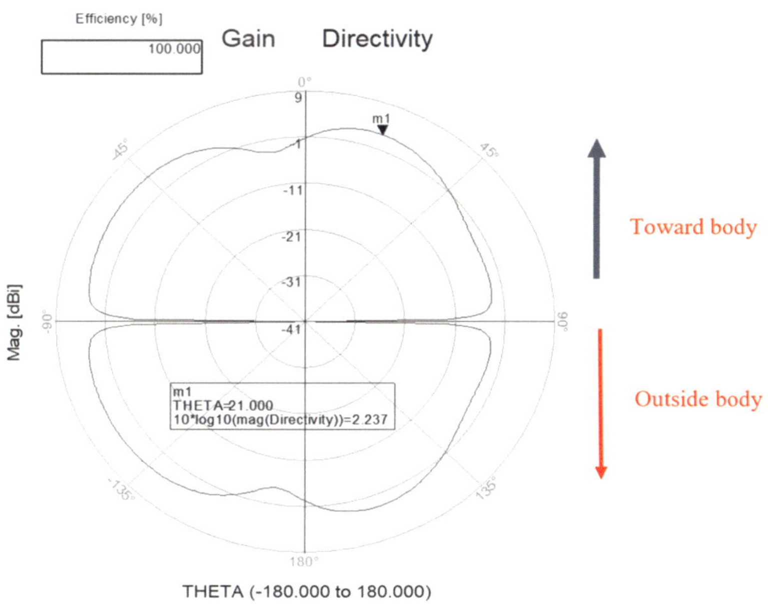


**Figure 4.**
*Radiation pattern of the notch antenna at 3.5 GHz.*

to 7.8 GHz. The antenna beam width is around 80° at 5 GHz. The antenna gain is around 2.5 dBi at 3 GHz. The notch antenna radiation pattern at 3.5 GHz is presented in **Figure 4**. The electromagnetic energy is converted to DC energy by a rectifying circuit connected to the antenna input. A rechargeable battery is connected to the output of the rectifying circuit.

## 4. New wideband active 0.5–3 GHz energy harvesting notch antenna

Harvested power from RF transmitting links is usually lower than 0.1 $\mu W/cm^2$. Active antennas may improve the energy.

Improve harvesting system efficiency. A wideband active notch antenna with fractal structure was printed on a 1.2 mm thick with dielectric constant of 2.2. The compact active notch antenna is shown in **Figure 5**. The notch antenna dimensions are 74.5 × 57.1 mm. The antenna center frequency is 1.75 GHz. The active antenna bandwidth is around 150–200% for VSWR better than 3:1. The notch antenna gain, S21 parameter, is presented in **Figure 6**. The active antenna gain is 23 ± 3 dB for frequencies from 200 to 900 MHz.

The active notch antenna VSWR is better than 3:1 for frequencies from 0.5 to 3 GHz. The antenna beam width is around 84° at 1 GHz. A compact E-PHEMT LNA, low-noise amplifier, is connected to the notch antenna via an input matching network. An output matching network connects the amplifier port to the rectifying circuit. A printed compact DC voltage bias network supplies the bias voltages to the harvesting system. The amplifier specification is listed in **Table 4**. The amplifier

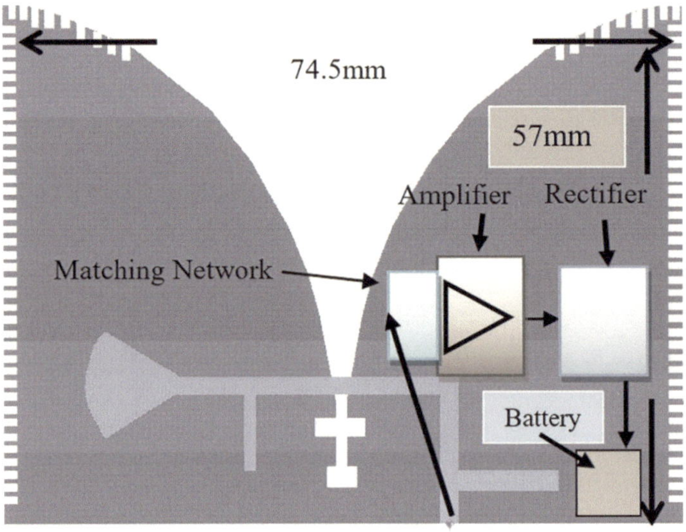


**Figure 5.**
*A wideband fractal active notch antenna.*

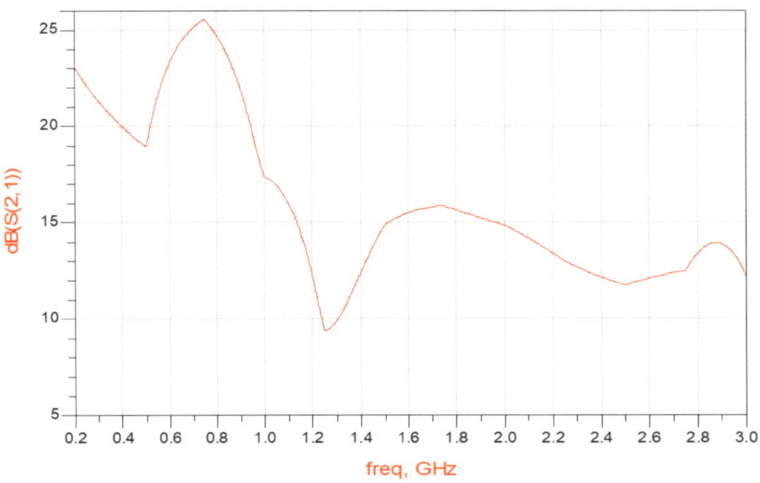


**Figure 6.**
*Active notch antenna S21 parameter.*

| Parameter | Specification | Remarks |
|---|---|---|
| Frequency range | 0.4–3 GHz | |
| Gain | 26 dB at 0.4 GHz<br>18 dB at 2 GHz | Vds = 3 V; Ids = 60 mA |
| Noise figure | 0.4 dB at 0.4 GHz<br>0.5 dB at 2 GHz | Vds = 3 V; Ids = 60 mA |
| P1dB | 18.9 dBm at 0.4 GHz<br>19.1 dBm at 2 GHz | Vds = 3 V; Ids = 60 mA |
| OIP3 | 32.1 dBm at 0.4 GHz<br>33.6 dBm at 2 GHz | Vds = 3 V; Ids = 60 mA |
| Max. input power | 17 dBm | |
| Vgs | 0.48 V | Vds = 3 V; Ids = 60 mA |
| Vds | 3 V | |
| Ids | 60 mA | |
| Supply voltage | ±5 V | |
| Operating temp. | −40–80°C | |

**Table 4.**
*LNA amplifier specification.*

complex S parameters are listed in **Tables 5** and **6**. The amplifier noise parameters are listed in **Table 7**.

The active antenna gain is 12 ± 3 dB for frequencies from 1 to 3 GHz. The active notch antenna noise figure is 0.5 ± 0.3 dB for frequencies ranging from 300 to 3.0 GHz as presented in **Figure 7**. The active notch antenna output VSWR is better than 3:1 for frequencies from 0.5 to 3 GHz. All antennas presented in this paper can operate as passive and active antennas.

| F-GHz | S11 | S11° | S21 | S21° |
|---|---|---|---|---|
| 0.19 | −31.76 | 0.964 | 24.13 | 158.9 |
| 0.279 | 0.93 | −45.77 | 22.97 | 149.5 |
| 0.323 | 0.92 | −53.39 | 22.45 | 145.3 |
| 0.413 | 0.89 | −65.72 | 20.98 | 137.27 |
| 0.50 | 0.87 | −77.1 | 19.54 | 130.3 |
| 0.59 | 0.83 | −87.12 | 18.08 | 124.14 |
| 0.726 | 0.8 | −100.8 | 16.22 | 115.7 |
| 0.816 | 0.77 | −108.8 | 15.07 | 110.75 |
| 1.04 | 0.74 | −126.2 | 12.74 | 100.13 |
| 1.21 | 0.71 | −137.6 | 11.25 | 92.91 |
| 1.53 | 0.687 | −154.2 | 9.29 | 82.06 |
| 1.75 | 0.67 | −164.1 | 8.24 | 75.31 |
| 2.02 | 0.67 | −174.6 | 7.27 | 67.82 |

**Table 5.**
*LNA amplifier S parameters.*

| F-GHz | S12 | S12° | S22 | S22° |
|---|---|---|---|---|
| 0.19 | 0.016 | 74.88 | 0.54 | −22.98 |
| 0.279 | 0.021 | 65.77 | 0.51 | −33.65 |
| 0.323 | 0.026 | 62.38 | 0.49 | −39.2 |
| 0.413 | 0.03 | 57.9 | 0.46 | −49.3 |
| 0.50 | 0.034 | 53.03 | 0.43 | −57.5 |
| 0.59 | 0.038 | 48.18 | 0.40 | −64.12 |
| 0.726 | 0.042 | 42.06 | 0.36 | −74.86 |
| 0.816 | 0.044 | 39.53 | 0.34 | −80.87 |
| 1.04 | 0.049 | 33.69 | 0.29 | −94.96 |
| 1.21 | 0.051 | 30.05 | 0.26 | −104 |
| 1.53 | 0.055 | 26.08 | 0.22 | −119 |
| 1.75 | 0.058 | 23.14 | 0.20 | −128.4 |
| 2.02 | 0.06 | 20.88 | 0.18 | −138.8 |

**Table 6.**
*LNA amplifier S parameters.*

| F-GHz | NFMIN | N11X | N11Y | Rn |
|---|---|---|---|---|
| 0.4 | 0.070 | 0.3276 | 20.05 | 0.062 |
| 0.5 | 0.079 | 0.3284 | 24.56 | 0.056 |
| 0.7 | 0.112 | 0.334 | 36.08 | 0.050 |
| 0.9 | 0.144 | 0.3396 | 47.4 | 0.045 |
| 1 | 0.16 | 0.3424 | 52.98 | 0.042 |
| 1.9 | 0.306 | 0.3682 | 100.93 | 0.029 |
| 2 | 0.322 | 0.3711 | 106.01 | 0.029 |
| 2.4 | 0.387 | 0.3829 | 125.79 | 0.029 |
| 3 | 0.484 | 0.401 | 153.93 | 0.036 |
| 3.9 | 0.629 | 0.429 | −167.3 | 0.059 |
| 5 | 0.808 | 0.4645 | −125.53 | 0.11 |

**Table 7.**
*LNA noise parameters.*

## 5. New fractal active 0.4–3 GHz energy harvesting antenna

A compact notch antenna with fractal structure is shown in **Figure 8**. The antenna is printed on a 5580 Duroid substrate, 1.2 mm thick, with dielectric constant of 2.2. The notch antenna dimensions are 52.2 × 36.8 × 1.2 mm. The antenna center frequency is 1.7 GHz. The antenna bandwidth is around 100% for VSWR better than 3:1. The active notch antenna VSWR is better than 3:1 for frequencies from 0.4 to 3 GHz. The antenna beam width is around 82° at 1 GHz. An LNA is connected to the antenna feed line. The antenna is connected to the LNA via an input matching network. An output matching network connects the amplifier output port to a rectifying circuit. A compact DC network supplies the required voltages to the active antenna. The amplifier specification, S parameters, and the

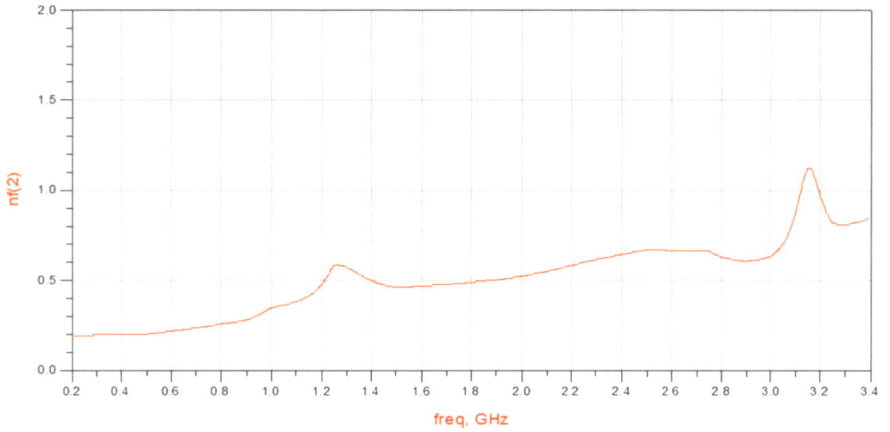


**Figure 7.**
*Active notch antenna noise figure.*

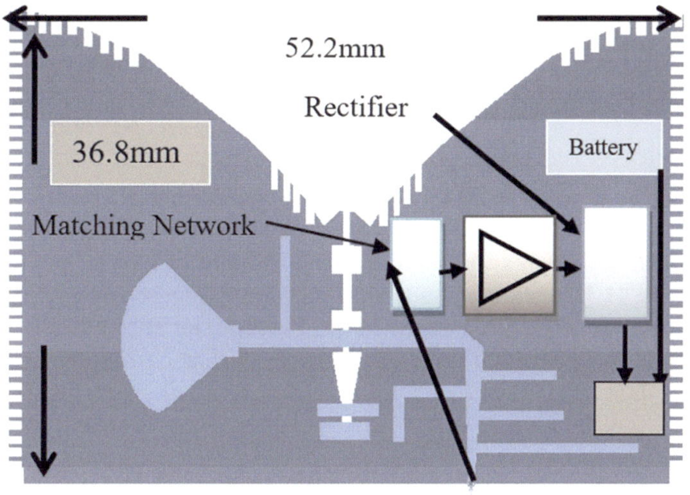


**Figure 8.**
*A wideband fractal active notch antenna with fractal structure.*

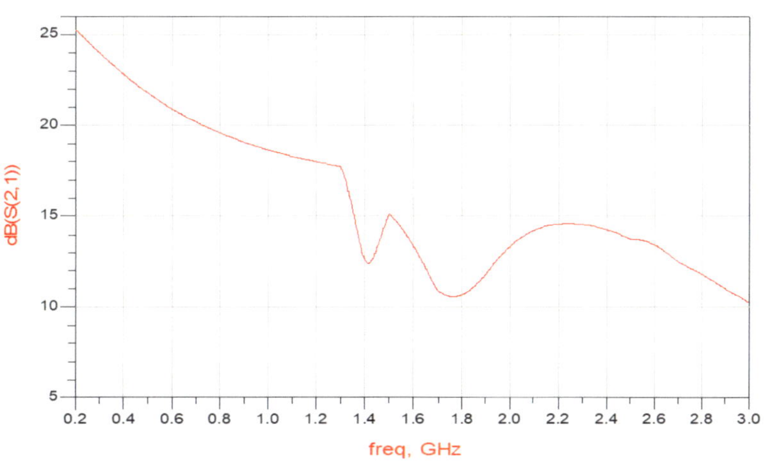


**Figure 9.**
*A fractal active notch antenna S21 parameter.*

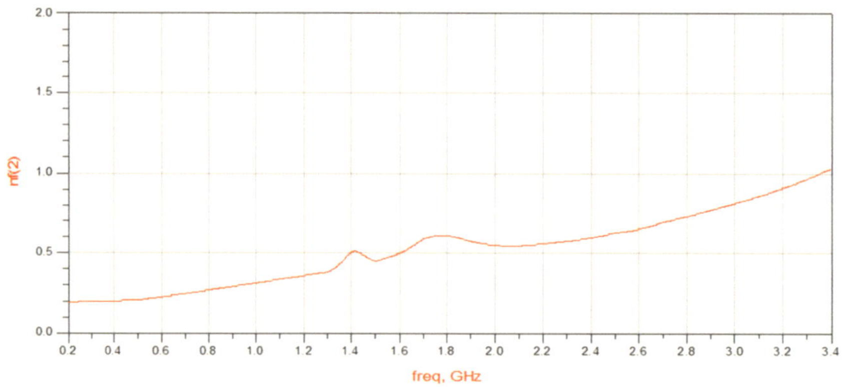


**Figure 10.**
*Active fractal notch antenna noise figure.*

amplifier noise parameters are listed in **Tables 4–7**. The active antenna gain is 21 ± 3 dB for frequencies from 400 MHz to 1.3 GHz, as presented in **Figure 9**. The active antenna gain is 12 ± 3 dB for frequencies from 1.3 to 3 GHz. The active notch antenna noise figure is 0.5 ± 0.3 dB for frequencies from 300 MHz to 3.0 GHz, as presented in **Figure 10**. The notch antenna output VSWR is better than 3:1 for frequencies from 0.5 to 3 GHz.

## 6. New wideband active 0.8–5.4 GHz energy harvesting slot antenna

A wideband T shape wearable slot antenna for energy harvesting applications is shown in **Figure 11**. The antenna electrical parameters were computed by using momentum software [38]. The volume of the T-shape slot antenna is 7 × 7 × 0.12 cm. The slot antenna center frequency is around 3 GHz. The computed S11 parameters

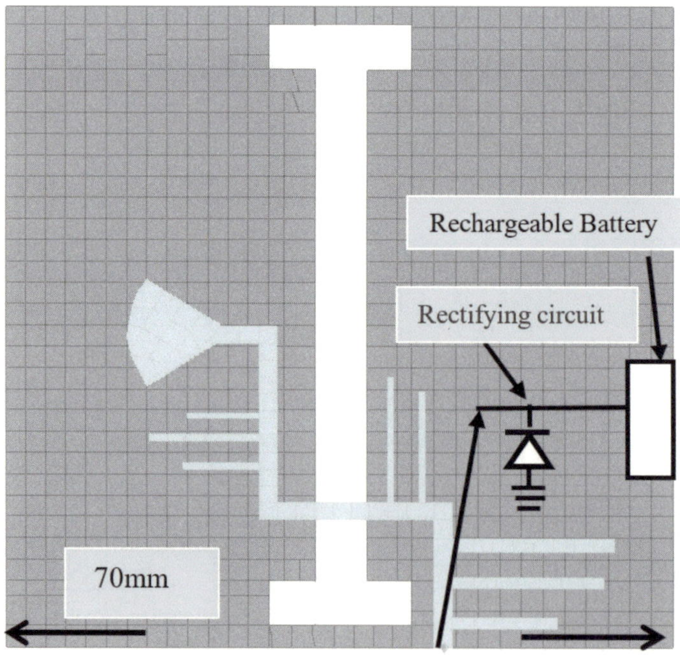


**Figure 11.**
*A wideband energy harvesting slot antenna.*

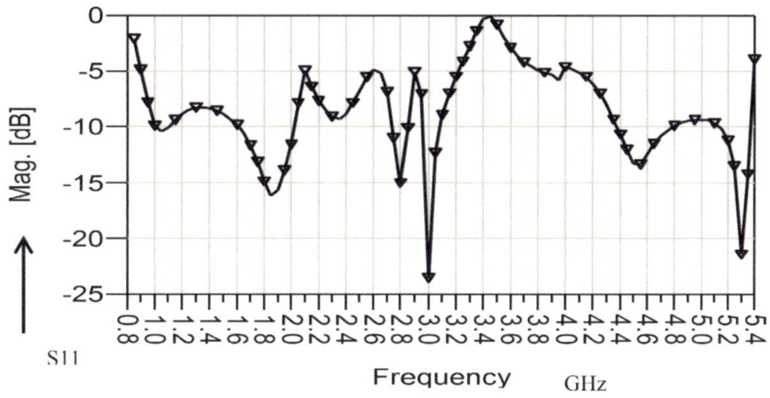


**Figure 12.**
*Computed S11 of a wideband, 0.8–5.4 GHz, slot.*

are presented in **Figure 12**. The antenna bandwidth is around 100% for VSWR better than 3:1. The antenna beamwidth is around 138° at 1 GHz as shown in **Figure 13**. The antenna gain is around 2.5 dBi. The antenna was designed also as an active antenna as shown in **Figure 14**. The slot antenna is connected to the LNA via an input matching network. The output matching network connects the amplifier output port to a rectifying circuit. For example, a MMIC LNA with 16 dB gain and 1 dB noise figure has DC power consumption of less than 18 mW in the frequency range from 70 MHz to 1 GHz. The system DC bias network supply the required voltages to the energy harvesting system. The amplifier specification is listed in **Table 5**. The amplifier complex S parameters are listed in **Tables 5** and **6**. The active slot antenna gain, S21 parameter, is presented in **Figure 15**. The active antenna gain is 23 ± 3 dB for frequencies from 200 to 900 MHz. The active antenna gain is 13 ± 3 dB for frequencies from 1 to 3 GHz. The active slot antenna noise

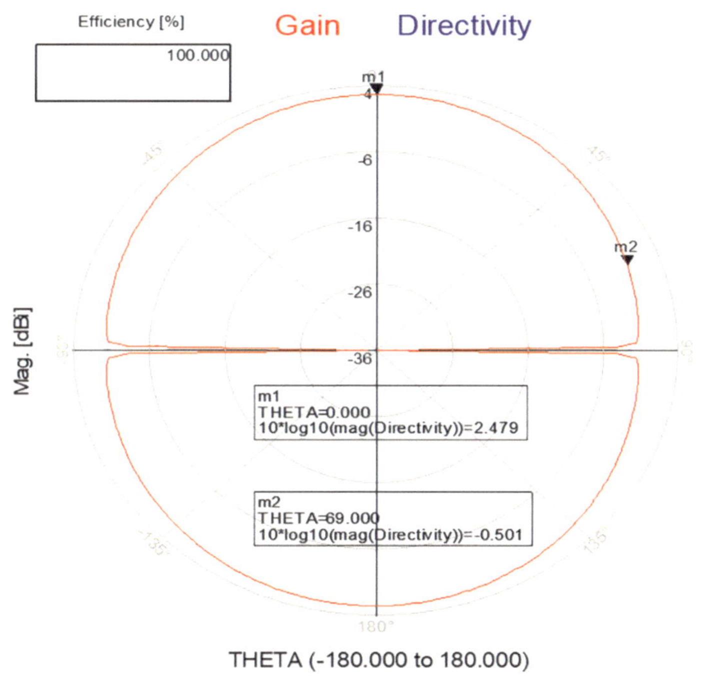


**Figure 13.**
*Radiation pattern of the energy harvesting slot antenna.*

figure is 0.5 ± 0.3 dB for frequencies from 200 MHz to 3.0 GHz. The computed S11 parameters of the T-shape slot on human body are presented in **Figure 16**. The dielectric constant of human body tissue was taken as 45. The antenna was attached to a shirt with dielectric constant of 2.21 mm thick.

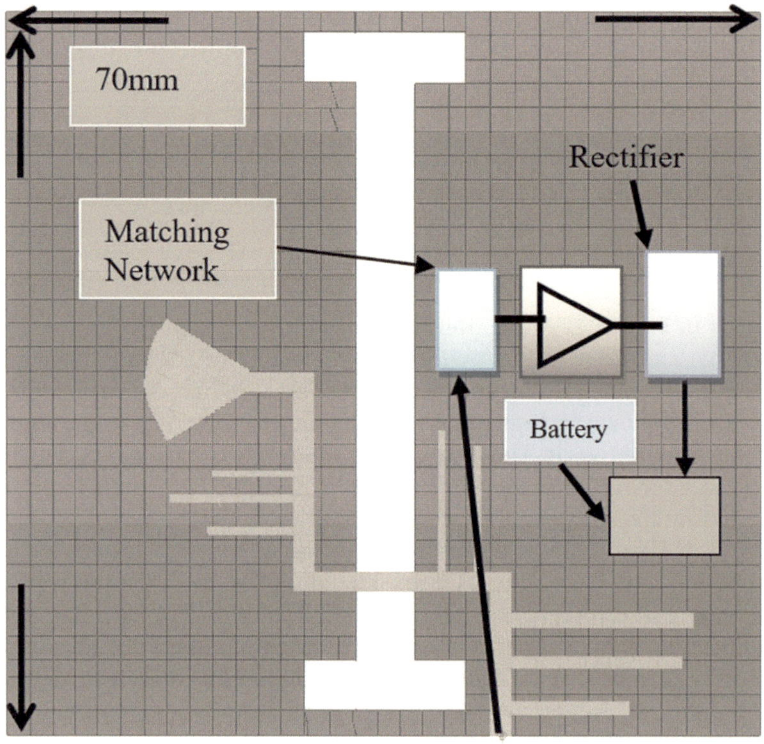


**Figure 14.**
*A wideband active energy harvesting slot antenna.*

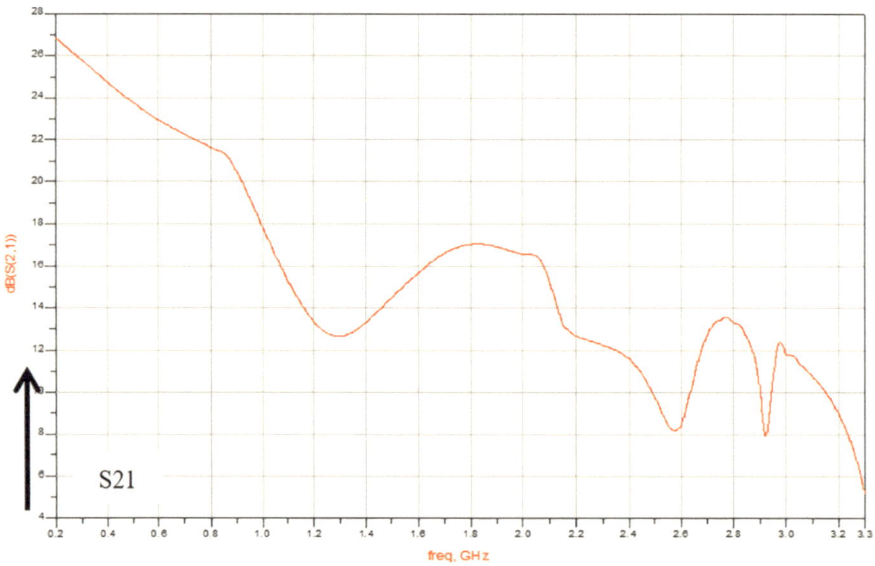


**Figure 15.**
*Active energy harvesting slot antenna S21.*

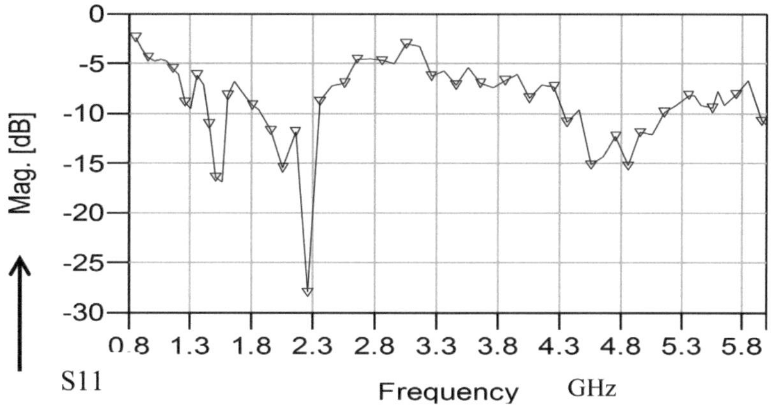


**Figure 16.**
*Computed S11 of a wideband, 0.8–5.4 GHz, slot. Antenna on the human body.*

## 7. Energy harvesting systems for medical and IOT applications

The notch and slot antennas' electrical performance near the human body was investigated by using the model shown in **Figure 17**. Properties of human body tissues are listed in **Table 8** [27, 28]. These properties were employed in the antenna design. Up to four energy harvesting antennas may be assembled in a belt and attached to the patient body as presented in **Figure 18**. The bias voltage to the active elements is supplied by a compact recorder battery. The DC cables from each harvesting antenna are connected to a rechargeable battery. The electromagnetic energy is converted to DC energy that may be employed to charge medical or commercial body area networks (BANs).

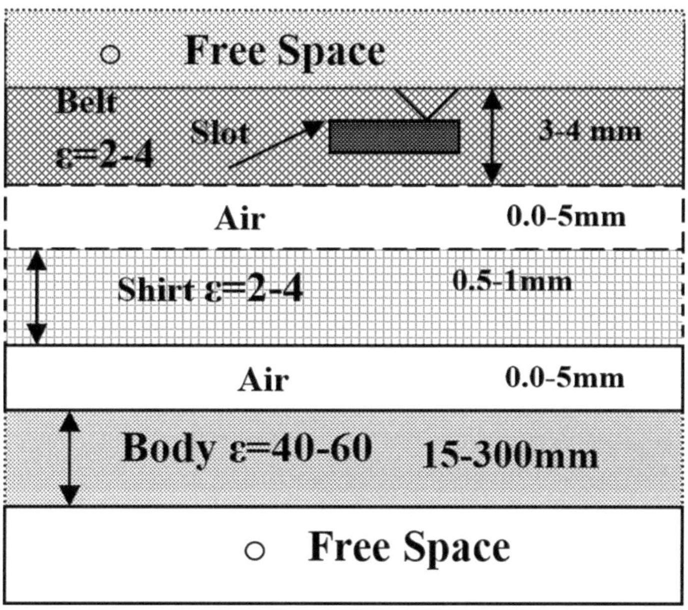


**Figure 17.**
*Analyzed structure for wearable slot antennas.*

| Tissue | Property | 600 MHz | 1000 MHz |
|---|---|---|---|
| Fat | σ | 0.05 | 0.06 |
| | ε | 5 | 4.52 |
| Stomach | σ | 0.73 | 0.97 |
| | ε | 41.41 | 39.06 |
| Colon | σ | 1.06 | 1.28 |
| | ε | 61.9 | 59.96 |
| Lung | σ | 0.27 | 0.27 |
| | ε | 38.4 | 38.4 |
| Prostate | σ | 0. 75 | 0.90 |
| | ε | 50.53 | 47.4 |
| Kidney | σ | 0.88 | 0.88 |
| | ε | 117.43 | 117.43 |

**Table 8**
*Electrical properties of human body tissues.*

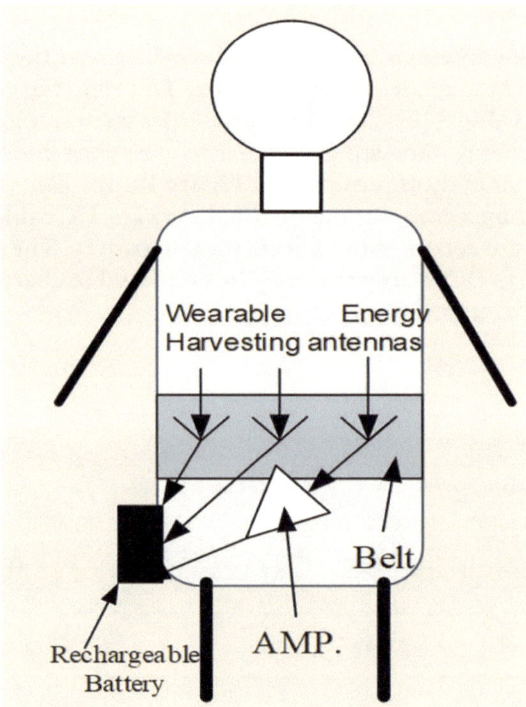


**Figure 18.**
*Active wearable energy harvesting antennas.*

## 8. Energy harvesting system

As presented in **Figure 1**, the energy harvesting system consists of an antenna, a rectifying circuit, and a rechargeable battery. A rectifier is a circuit that converts electromagnetic energy, alternating current AC, to direct current (DC). Half-wave rectifier or full wave rectifier may be used to convert electromagnetic AC energy to DC electrical energy. A half-wave rectifier is presented in **Figure 19**. A half-wave rectifier conducts only during the positive half cycle. It allows only one half of an AC waveform to pass through the load. The rectifier output DC voltage, $V_{ODC}$, is

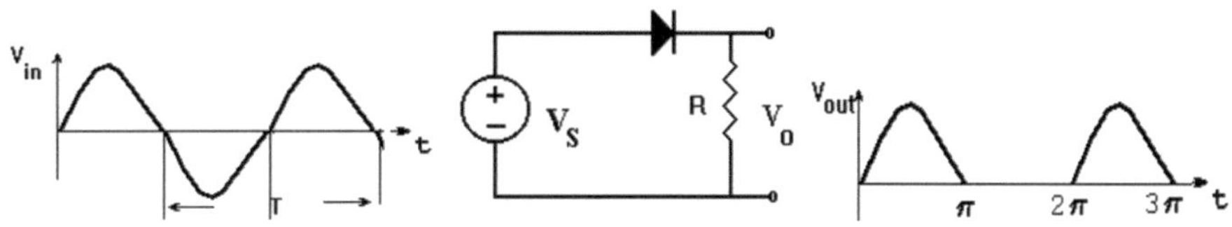


**Figure 19.**
*Half-wave rectifier.*

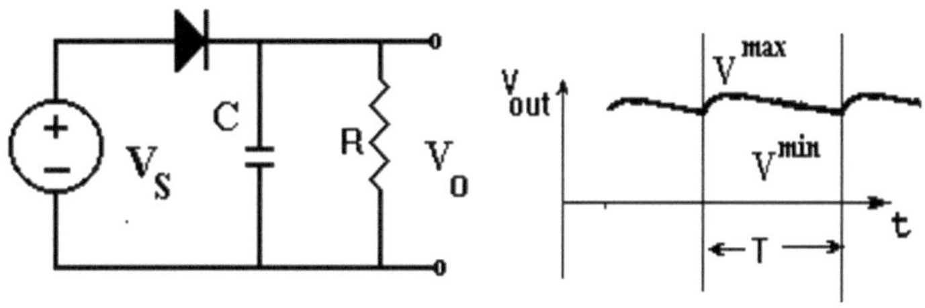


**Figure 20.**
*Improved half-wave rectifier.*

given in Eq. 5. The rectifier output voltage may be improved by connecting a capacitor in shunt to the resistor as presented in **Figure 20**.

$$V_{O,DC} = \frac{1}{2\pi}\int_{0}^{2\pi} V_{O}^{MAX}\sin(\omega t)d(\omega t);\ \omega = 2\pi f$$
$$V_O = V_S - V_{DON} \approx V_S;\ V_O^{MAX} = V_m$$
$$V_{ODC} = V_m/\pi \tag{5}$$

The improved half wave rectifier is shown in **Figure 20**.

$$Vripple = Vr = Vmax - Vmin = {}^{VDC}/_{fCR} \tag{6}$$

The time constant $\tau$ should be lower than T, where, $\tau = RC \ll T$. The half-wave rectifier efficiency is 40.6% as given in Eq. 7. Only 40.6% of the input AC power is converted into DC power. Where rf the diode resistance is negligible as compared to R.

$$\eta = \frac{\text{DC output power}}{\text{AC input power}} = \frac{\left(\frac{I_m}{\pi}\right)^2 R}{\left(\frac{I_m}{2}\right)^2 (R + rf)} \sim 0.406 \tag{7}$$

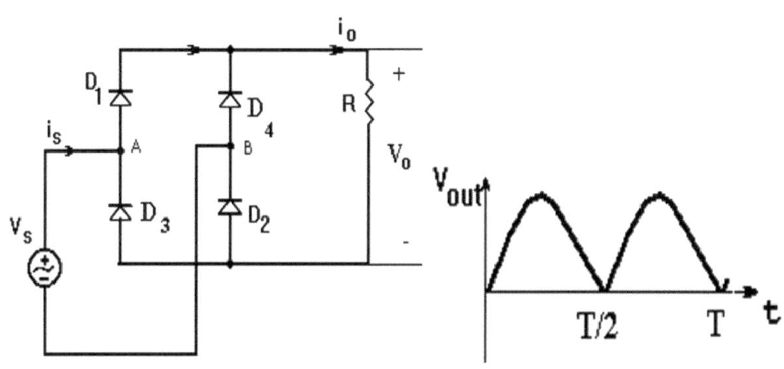


**Figure 21.**
*Full-wave rectifier.*

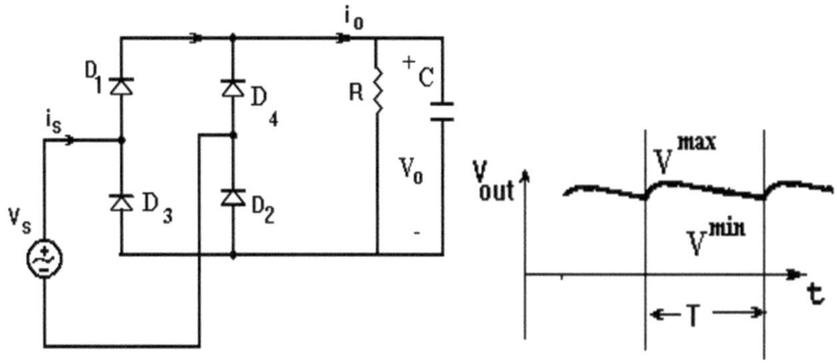


**Figure 22.**
*Improved full-wave rectifier.*

The bridge full-wave rectifier circuit is used usually for DC power supplies. It consists of four diodes, D1–D4, as shown in in **Figure 21**, connected to form a bridge. During the positive input half cycle, terminal A will be positive and terminal B will be negative. Diodes D1 and D2 will become forward biased and D3 and D4 will be reversed biased. The rectifier output DC voltage, $V_{ODC} = 2V_m/\pi$, may be improved by connecting a capacitor in the shunt to the resistor. The improved half-wave rectifier is presented in **Figure 22**.

The half-wave rectifier efficiency is 81.2% as presented in Eq. 8. This means that only 81.2% of the input AC power is converted into DC power.

$$\eta = \frac{\text{DC output power}}{\text{AC input power}} = \frac{\left(\frac{2I_m}{\pi}\right)^2 R}{\left(\frac{I_m}{2}\right)^2 (R + rf)} \sim 0.812 \tag{8}$$

A capacitor is used in the improved rectifier to get flat output voltage variation as function of time. The capacitor may be a voltage-controlled varactor diode. Varactors are voltage variable capacitors designed to provide electronic tuning of electrical devices. The output voltage ripple (see Eq. 6) of the improved rectifier may be tuned as function of the frequency of the received signal or of the load resistance R.

A Schottky diode may be used in the rectifier circuit. Schottky diodes are semi-conductor diodes which has a low forward voltage drop and a very fast switching action. There is a small voltage drop across the diode terminals when current flows through the diode. The voltage drop of a Schottky diode is usually between 0.15 and 0.4 V. This lower voltage drop provides better system efficiency and higher switching speed. A normal diode has a voltage drop between 0.6 and 1.7 V. Electrical characteristics of Schottky diode and standard PN diodes are listed in **Table 9**. Typical I–V curves of commercial Schottky diodes are shown in **Figure 23**.

| Parameter | Schottky diode | PN diode |
|---|---|---|
| Forward current mechanism | Majority carrier transport | Minority carrier transport |
| Reverse current | Less temperature dependence | Strong temperature dependence |
| Turn on voltage | Small—around 0.2 V | Comparatively large around 0.7 V |
| Switching speed | Fast | Limited |

**Table 9.**
*Electrical characteristics of Schottky diode and PN diodes.*

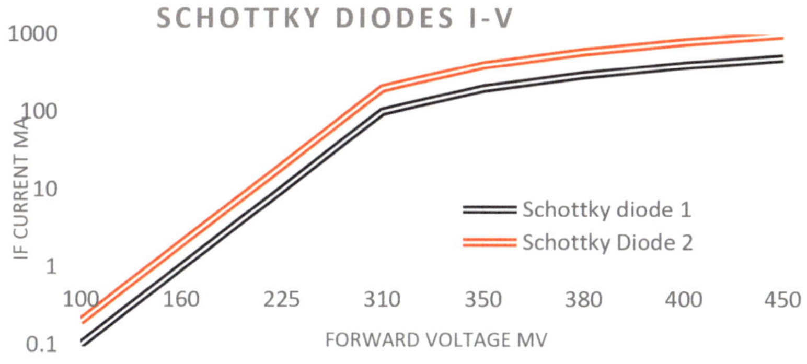


**Figure 23.**
*Typical I–V curves of Schottky diodes.*

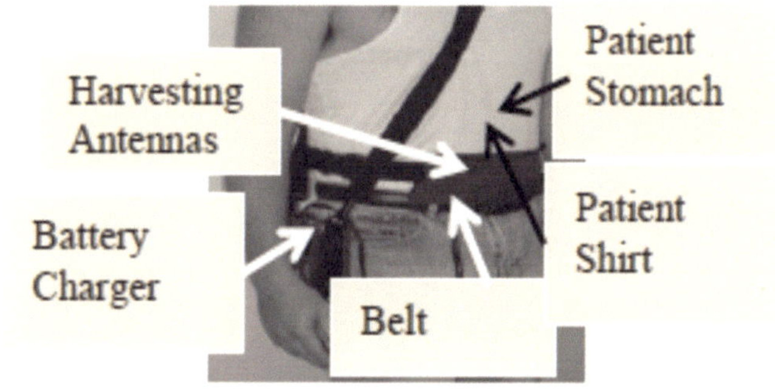


**Figure 24.**
*Medical wearable harvesting system.*

**Figure 24** presents a wearable harvesting system and a wearable battery charger attached to the patient shirt.

## 9. Conclusion

This paper presents new ultra-wideband wearable passive and active energy harvesting systems and antennas in frequencies ranging from 0.4 to 8 GHz. The antennas are inserted in a belt and attached to the body. The antennas are compact and can be attached to the body. The antennas allow the patients' easy movement (running, jumping, and working).

The electromagnetic energy is converted to DC energy that may be employed to charge batteries, wearable medical devices, and commercial body area networks. The passive and active notch and slot antennas were analyzed by using 3D full-wave software. Harvested power from RF transmitting links is usually lower than 0.1 μW/ $cm^2$. Active antennas may improve the energy harvesting system efficiency. All antennas presented in this paper can operate also as passive antennas. The active notch and slot antenna bandwidth are from 50 to 100% with VSWR better than 3:1. The slot antenna gain is around 3 dBi with efficiency higher than 90%. The antenna electrical parameters were computed near the human body. The active slot antenna gain is 24 $\pm$ 2.5 dB for frequencies ranging from 200 to 900 MHz. The active slot antenna gain is 13 $\pm$ 3dB for frequencies from 1 to 3.3 GHz. The active wearable antennas may be used in energy harvesting systems for wireless communication and medical applications. The RF energy harvesting system consists of an antenna, a rectifying circuit, and a rechargeable battery. The harvesting energy system operates as a dual mode energy harvesting system. The low-noise amplifier is part of the receiving system. The LNA DC bias voltages are supplied by the receiving system.

## Author details

Albert Sabban[1,2]

1 Kinneret Academic College, Israel

2 Ort Braude College, Karmiel, Israel

*Address all correspondence to: sabban@mx.kinneret.ac.il

## References

[1] Paradiso JA, Starner T. Energy scavenging for mobile and wireless electronics. IEEE Pervasive Computing. 2005;**4**(1):18-27

[2] Valenta CR, Durgin GD. Harvesting wireless power: Survey of energy-harvester conversion efficiency in far-field, wireless power transfer systems. IEEE Microwave Magazine. 2014;**15**(4): 108-120

[3] Nintanavongsa P, Muncuk U, Lewis DR, Chowdhury KR. Design optimization and implementation for RF energy harvesting circuits. IEEE Journal on Emerging and Selected Topics in Circuits and Systems. 2012; **2**(1):24-33

[4] Devi KKA, Sadasivam S, Din NM, Chakrabarthy CK. Design of a 377 Ω patch antenna for ambient RF energy harvesting at downlink frequency of GSM 900. In: Proceedings of the 17th Asia Pacific Conference on Communications (APCC '11). Malaysia: Sabah; October 2011. pp. 492-495

[5] Rahim RA, Malek F, Anwar SFW, Hassan SLS, Junita MN, Hassan HF. A harmonic suppression circularly polarized patch antenna for an RF ambient energy harvesting system. In: Proceedings of the IEEE Conference on Clean Energy and Technology (CEAT '13). Malaysia: IEEE, Lankgkawi; November 2013. pp. 33-37

[6] Krakauskas M, Sabaawi AMA, Tsimenidis CC. Suspended patch microstrip antenna with cut rectangular slots for RF energy harvesting. In: Proceedings of the 10th Loughborough Antennas and Propagation Conference (LAPC '14). UK: Loughborough; November 2014. pp. 304-307

[7] Sabban A. Low Visibility Antennas for Communication Systems. USA: Taylor & Francis group; 2015

[8] Sabban A. Wideband RF Technologies and Antenna in Microwave Frequencies. USA: Wiley Sons; July 2016

[9] James JR, Hall PS, Wood C. Microstrip Antenna Theory and Design. London; 1981

[10] Sabban A, Gupta KC. Characterization of radiation loss from microstrip discontinuities using a multiport network modeling approach. IEEE Transactions on Microwave Theory and Techniques. April 1991; **39**(4):705-712

[11] Sabban A. A new wideband stacked microstrip antenna. In: IEEE Antenna and Propagation Symp., Houston, Texas, USA. June 1983

[12] Sabban A, Navon E. A MM-waves microstrip antenna array. In: I.E.E.E Symposium, Tel-Aviv. March 1983

[13] Balanis CA. Antenna Theory: Analysis and Design. 2nd ed. Wiley; 1996

[14] Sabban A. Multiport network model for evaluating radiation loss and coupling among discontinuities in microstrip circuits [PhD thesis]. University of Colorado at Boulder; January 1991

[15] Sabban A. Microstrip antenna arrays. U.S. Patent US 1986/4,623,893; 1986

[16] Sabban A. Wideband microstrip antenna arrays. I.E.E.E Antenna and Propagation Symposium MELCOM, Tel-Aviv; June 1981

[17] Fujimoto K, James JR, editors. Mobile Antenna Systems Handbook. Boston, USA: Artech House; 1994

[18] Sabban A. New wideband notch antennas for communication systems.

Wireless Engineering and Technology Journal. April 2016:75-82

[19] Sabban A. Dual polarized dipole wearable antenna. U.S Patent number: 8203497. USA, June 19, 2012

[20] Sabban A. Wideband tunable printed antennas for medical applications. In: IEEE Antenna and Propagation Symp., Chicago, IL, USA. 2012, July 2012

[21] Sabban A. New wideband printed antennas for medical applications. I.E.E.E Transactions on Antennas and Propagation, January. 2013;**61**(1):84-91

[22] Sabban A. Comprehensive study of printed antennas on human body for medical applications. International Journal of Advance in Medical Science (AMS). February 2013;**1**:1-10

[23] Kastner R, Heyman E, Sabban A. Spectral domain iterative analysis of single and double-layered microstrip antennas using the conjugate gradient algorithm. I.E.E.E Transactions on Antennas and Propagation. Sept. 1988; **36**(9):1204-1212

[24] Sabban A. Microstrip antenna arrays. In: Nasimuddin N, editor. Microstrip Antennas. InTech; 2011. pp. 361-384. Available at: http://www.intechopen.com/articles/show/title/microstrip-antenna-arrays. ISBN: 978-953-307-247-0

[25] Sabban A. New wideband wearable notch antenna for energy harvesting applications. In: 12th European Conference on Antennas and Propagation EuCAP 2018, United Kingdom. 2018

[26] Chirwa LC, Hammond PA, Roy S, Cumming DRS. Electromagnetic radiation from ingested sources in the human intestine between 150 MHz and 1.2 GHz. IEEE Transaction on Biomedical Engineering. April 2003; **50**(4):484-492

[27] Werber D, Schwentner A, Biebl EM. Investigation of RF transmission properties of human tissues. Advances in Radio Science. 2006;**4**:357-360

[28] Gupta B, Sankaralingam S, Dhar S. Development of wearable and implantable antennas in the last decade. In: Microwave Symposium (MMS), 2010 Mediterranean. 2010. pp. 251-267

[29] Thalmann T, Popovic Z, Notaros BM, Mosig JR. Investigation and design of a multi-band wearable antenna. In: 3rd European Conference on Antennas and Propagation, EuCAP. 2009. pp. 462-465

[30] Salonen P, Rahmat-Samii Y, Kivikoski M. Wearable antennas in the vicinity of human body. In: IEEE Antennas and Propagation Society International Symposium. Vol. 1. 2004. pp. 467-470

[31] Kellomaki T, Heikkinen J, Kivikoski M. Wearable antennas for FM reception. In: First European Conference on Antennas and Propagation, EuCAP. 2006. pp. 1-6

[32] Sabban A. Wideband printed antennas for medical applications. In: APMC 2009 Conference, Singapore. Dec. 2009

[33] Alomainy A, Sani A, et al. Transient characteristics of wearable antennas and radio propagation channels for ultrawideband body-centric wireless communication. IEEE Transactions on Antennas and Propagation. April 2009; **57**(4):875-884

[34] Klemm M, Troester G. Textile UWB antenna for wireless body area networks. I.E.E.E. Transactions on Antennas and Propagation. Nov. 2006; **54**(11):3192-3197

[35] Izdebski PM, Rajagoplan H, Rahmat-Sami Y. Conformal ingestible capsule antenna: A novel chandelier meandered design. IEEE Transactions on Antennas and Propagation. April 2009;**57**(4):900-909

[36] ADS software. Agilent. Available at: http://www.home.agilent.com/agilent/product.jspx?cc=IL&lc=eng&ckey=1297113&nid=34346.0.00&id=1297113

Chapter 3

# Optimum Efficiency Analysis of Ecofriendly WDM-POF Optical Coupler

*Hadi Guna, Mohammad Syuhaimi Ab-Rahman, Norhana Arsad, Roslan Shukor and Sahbudin Shaari*

## Abstract

This chapter was presented to promote the development of Wavelength Division Multiplexing, WDM networking system based on Polymer Optical Fiber (POF). 1 × 3 POF coupler has been fully utilized to couple the WDM optical signals. An optimum efficiency analysis and mathematical modeling has been conducted to produce an effective device to be integrated into WDM-POF system. But despite its high speed data transmission feature, optical fiber technology remains an expensive option in optical network, although the installation costs associated with fiber through the transmission network can be minimized through the fabrication of 1 × 3 POF coupler. The objective of this chapter is identifying the differences in factors affecting the output power of the POF 1 × 3 coupling. It is then able to develop an efficient WDM-POF network system with high output power at a minimal cost. Several measurements using a power meter record performance and analysis device losses and power outputs. The demultiplexer efficiency is approximately 70%. Demultiplexer fabrication is easy with color and epoxy filters, although for some parts it requires careful attention. The output shows that Ecofriendly WDM-POF Optical Coupler can be used as one of the low-cost media for home networking and automotive applications.

**Keywords:** PMMA, optical fiber, POF, WDM, network

## 1. Introduction

The research in this chapter was carried out to optimize the efficiency of components and the entire system of Wavelength Division Multiplexing, WDM networking system based on Polymer Optical Fiber (POF). To achieve this purpose, a deeper analysis was performed on 1 × 3 POF coupler using fusion techniques in order to obtain and observe a number of parameters which directly affect the value of coupler's efficiency.

POF has many advantages compared to silica fiber and copper wires for short-distance applications, such as low cost, easy installation and connection, the integration of low cost LED and so on. These advantages lead to a high demand on data transmission in various applications, especially regarding home-networking and In-Vehicle Infotainment, or IVI's system on automotive field.

POF was initially used in automotive application for the first time in 1998 by Daimler-Chrysler. The development of optical data path is known as Domestic Digital Bus (D2B). Sensing is just one application area in automotive where photonics is used [1]. Typically, transport data path protocol so-called Media Oriented System Transport (MOST) with a speed of 24.8 Mbit/sec is used in many cars. However, this transmission data with Time Division Multiplexing (TDM)'s based network (as shown in **Figure 1**) has its own disadvantages where in the event of one or more malfunction lines (in the ring topology) will affect other line or more in MOST system.

In this analysis we aim to provide a better solution to a problem in the topology of ring bandwidth that many foreign carmakers use for the infotainment system of their network.

Laser can be very hazardous as best transmitting medium that can be used alongside GOF until the leakage taken place from the GOF body structure. A fiber-released extremely high intensity light ray can potentially burn a human retinas and lead to a permanent blind. It was hard to imagine if this silica-based technology was put in place inside the vehicle where consumers used this data transmission services directly.

In the meantime, the combination of POF, which is very suitable for a light-emitting diode (LED) system, can be seen as the best solution to offer a more secure data communication network, not to mention the lowest cost we can get for initial and production costs. For applications such as machine or peripheral connections, control and monitoring, board interconnections and even domestic hi-fi systems, POF links are becoming increasingly popular. Unlike GOF, POF remains versatile with a large core diameter and low numerical aperture, resulting in a high capacity that they can bring along the fiber.

The '*ecofriendly*' area involves a rapidly changing group of methods and materials, ranging from energy generation techniques to non-toxic cleaning products. The general perception is that in recent decades the IT-revolution is bringing innovation and improvements to daily life of a similar scale. In these early stages, it is difficult to predict what may potentially constitute '*ecofriendly*' [2].

In the current world, IT-societies face an increasingly serious challenge: on the one hand, the multimedia-rich data transmitted travels at amazing speed, and on the other the total energy consumption of communication and networking devices and the resulting global $CO_2$ emission are rising at an enormous rate. It is reported that currently, 3% of the world's energy is used by ICT, a system that generates roughly 2% of the world's $CO_2$ emissions, equal to worldwide $CO_2$ emissions by aircraft or a fifth of the world's $CO_2$ emissions by vehicle [3].

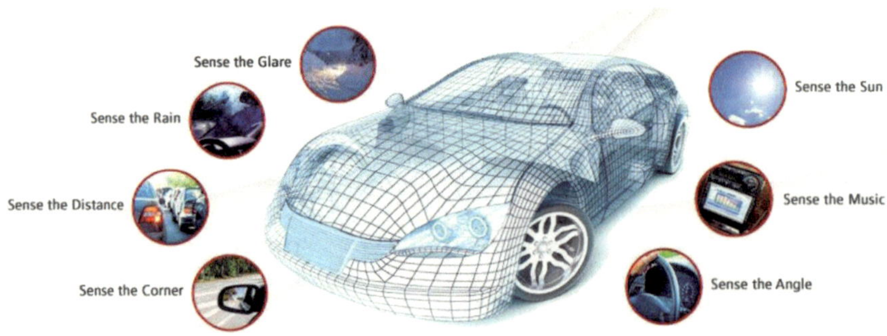


**Figure 1.**
*Sensor application using photonics in automotive area (source: hamamatsu.com).*

According to Ericsson Media Relations' recent research survey, cost of energy account for roughly half the operating costs of a mobile operator. Telecommunications technologies may therefore have a clear, tangible impact on reducing greenhouse gas emissions, power consumption, and effectively recycling infrastructure waste [4].

Therefore, finding solutions for optical communication's system that can greatly improve energy and resource efficiency not only helps the global environment, but also makes economic sense for telecom operators promoting viable and profitable business. Within the context for '*ecofriendly*' a range of paradigm shifting technical solutions can be anticipated, including but not limited to energy-efficient network architecture & protocols, energy-efficient wireless transmission technologies (*e.g.* reduced transmission capacity & radiation), cross-layer synchronization strategies, and opportunistic spectrum exchange without creating dangerous inter-radiation [4].

The Ecofriendly WDM-POF is provided on the basis of a network system consisting from POF. Ecofriendly WDM-POF is developed using a method that is environmentally friendly to divide and recombine a range of wavelengths. The Ethernet connection, DVD player and CCTV system was extensively used in three different wavelengths from ecologically friendly LED transmission systems. Red LED that is able to download and upload information over a 650 nm wavelength via Ethernet cable and green LED sends a video signal at a wavelength of 520 nm for DVD player while 470 nm LED will distribute the captured video using CCTV system. The Ecofriendly WDM-POF system will select a specific signal and produce it, if required. Special filter has been placed between the coupler and the receiving point. System and network performance are observed. In this chapter the product, production process, device and implementation strategy is focused on an environmentally sustainable solution for reducing energy usage and waste without impacting overall performance. The first documented approach in this paper is our Ecofriendly WDM-POF network implementation.

## 2. The advantages of LED lights for the environment

Taking care of the environment is a duty that everyone should feel responsible for. Most of us already know of environmentally friendly practices such as recycling to minimize pollution and reduce our carbon footprint. Most citizens, though, do not learn of new and upcoming technology that we can use to help reduce carbon emissions. Light-emitting diode (LED) lighting is a good example of this, which brings some environmental benefits. LED is a light source with a semiconductor that emits light when the current passes through it. Electrons, recombine with electron holes in the semiconductor, releasing energy in photon form. The color of the light (corresponding to the photons' energy) is determined by the energy that electrons need to reach the semiconductor's band gap. White light is generated on the semiconductor device using several semiconductors or a film of light-emitting phosphorus.

LED lighting is up to 80% more effective than conventional lighting including fluorescent or incandescent lamps. 95% of energy is turned to light in LEDs and only 5% is lost as heat. This is contrasted to fluorescent lights that turn 95% of energy to heat and only 5% to light LED lights can have much less power than conventional lighting; a traditional 84 watt fluorescent can be replaced with a 36 watt LED that offers the same amount of illumination. Lower energy consumption reduces demand from power plants and lowers emissions of greenhouse gases.

LED lights do not contain toxic materials. Many workplaces still use fluorescent strip lamps containing harmful chemicals like mercury. It contaminates the atmosphere when disposed of in waste pits. Disposal needs to be managed via a certified waste carrier so that switching to LED reduces costs and time to enforcement–which helps protect the environment against further toxic waste.

LEDs have a better light transmission efficiency and focus light in one direction, compared to other types of light that waste energy in all directions, often illuminate areas where no light is needed (*e.g.* the ceiling). It needs fewer LED lights to achieve the same level of luminosity as fluorescent and incandescent lamps. Less lighting can reduce energy usage and thus benefit the environment.

A longer life span means lower production of carbon LED lights last up to six times longer than other lamp types, eliminating the regular maintenance demand. This leads to the use of fewer lamps and therefore fewer resources for production, processing and transport processes [5].

## 3. A novel fused coupler

A novel fused coupler and the special filter –an advanced design of WDM-POF network implementation, are two key elements of the WDM-POF system in this chapter. We suggest an innovative fusion technology that is simple and inexpensive to use with Bunsen burner and metal tube to produce indirect heating processes with low structural defects, low excess losses and a good splitting ratio in the production of POF-based coupler.

The term '*fusion*' typically describes the operation or method of liquefying and melting by applying heat. The fabrication approach varies notably from traditional biconical technique. Seeing that the fabrication of polymer fiber is not subject to very high temperatures, POF's are liquefied with yellow flame Bunsen burners (1000°C) indirect heat treatment rather than with an oxyhydrogen burner (heating temperatures T = 2660°C) applied conventionally to the production of GOF-based couplers [6, 7].

Metal tube is used to protect the POF structure from direct heater during the fiber fusion process of indirect heating. POFs which are vulnerable to severe damage in their core cause heating a bundle of POFs directly to the burner flame.

In the new method of fusion, as shown in **Figure 2**, fusion length $L_f$, twist $T$ number, fusion time $t_f$, pulling length $L_p$ and other parameters are controllable parameters. The multimode step indexed POF with a core diameter of 1 mm of *Polymethylmethacrylate* (PMMA) was used as material for second-generation couplers. PMMA is one of the optical components most frequently used. The inherent absorption loss is mainly due to the stretching vibration of carbon hydrogen in PMMA core [7]. *Polyvinylchloride* (PVC) is another component used to cover the POF ports as a jacket.

The high cost of industrial coupler was posed as a crucial obstacle for the creation of wavelength division multiplexing (WDM), according to Kagami's 2006 Toyota R&D analysis report [8]. The fabrication process of expensive BFT device is considered a high production cost factor. The diameter of the fused taper area is extremely small by traditional BFT fabrication process, in which strain is accumulated and result in poor structural survivability.

Following the adaptation of the fused tapering technique for traditional multimode fiber, we successfully established the fabrication process for fused taper couplers with 1 × N POF. The handmade 1 × N coupler is an optical device that ends with *N* number of POF output terminals, while the other ends with one POF port. Like other traditional couplers, bidirectional operation is also possible, working from the *N* ports to 1 port (for coupling signal purposes) or vice versa (for splitting signals purposes). For instance, the optical 1 × 4 coupler formed by combining four *Polymethylmethacrylate* (PMMA) POF [9]. The output

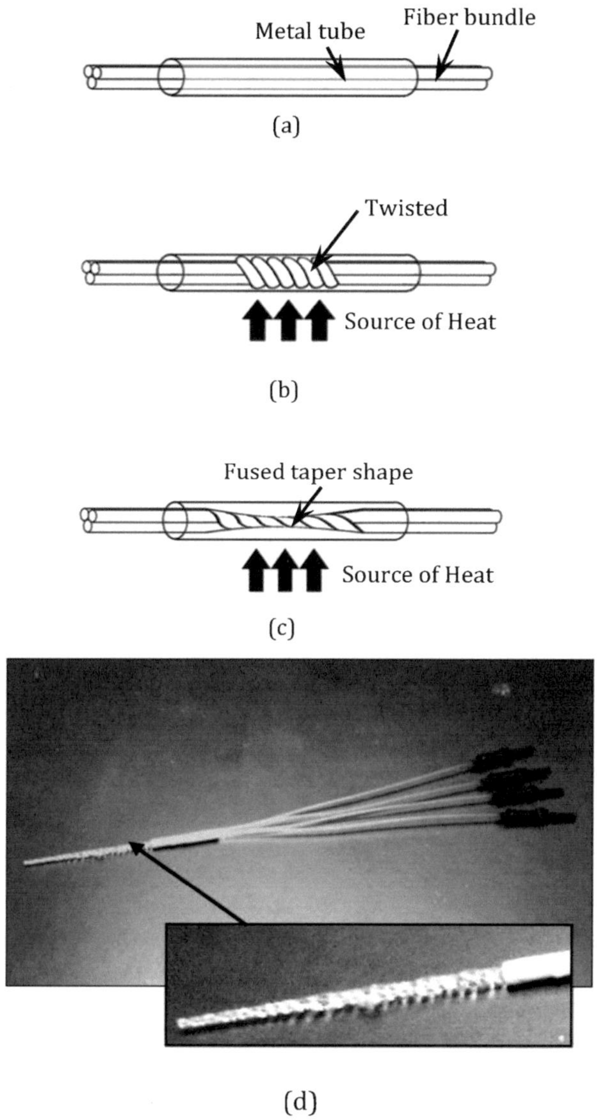


**Figure 2.**
*Fabrication process of* fused tapered *coupler: (a) configuration, (b) twisting and (c) fusion and (d) prototype of the 1 × 4 handmade coupler fabricated by fused tapering technique.*

POF is designed and fabricated to fuse the tapered shape in some detail. **Figure 2** shows the process 1 × 4 POF coupler.

Standard multimode SI-POF is used with its core diameter of 980 μm, cladding thickness of 10 μm and the refractive index is 1.49. To obtain the results, demultiplexer is realized using handmade color films attached using epoxy resin to the edge of the connectors. The components are chosen because they are low cost and are easily found in the market [10].

## 4. Ecofriendly WDM-POF

For applications like home networking, car industry, board interconnection, also residential hi-fi networks POF connectivity is increasingly prevalent. POF is versatile, as compared to GOF, with a wide core and a lower numerical aperture [1] and it can carry a high fiber capacity The optical coupler plays an important role among passive components for POF technologies, which is enabled by the versatility of a full product range. A POF coupler has been developed with various techniques. Thermal deformation, refining and merging, hand polished, scraping, etching, molding, reflecting body and biconic body are all common techniques have been applied in recent years [11].

In order to increase bandwidth to wavelength division multiplexing (WDM) in beginning topology, our current POF communication system enables data to be conveyed over more than one single wavelength. WDM is a system in which several signals are transmitted together in a multiplexed signal as separate wavelengths of light. As the **Figure 3** shows, WDM Multiplexer is the first active WDM-POF network designed to merge optical signals onto a single fiber from multiple single-wave end devices. If one of the signals (audio, video, Ethernet, *etc.*) breaks down, the network will not impact others if the primary transmission line is ineffective.

In WDM-POF system, many transmitters with different lights color to carry single information. For example, red light with 665 nm wavelength modulated with Ethernet signal while blue ($\lambda_1$), green ($\lambda_2$) and yellow ($\lambda_3$) lights carry image information, RF and TV signal, respectively [7]. The Multiplexer (MUX) must couple the light and separate it by demultiplexer (DEMUX).

Due to its simple device approach for extending, WDM has extended over the past 20 years, the average bit rate of transmission into GOF-long-range systems: introducing another source of specific transmission wavelengths in combination with the MUX/DEMUX component directly increases the utilizable speed. The wavelengths for the WDM from 400 to 700 nm, as shown in **Figure 4**, are used because of the attenuation in POF.

By practice, the same device can also conduct the reverse of the same WDM methods, in which several wavelength information streams are split up into multiple single wavelength data flows. The reverse is termed de-multiplexing. Conceptually, as a single coupled signal, POF coupler has the same purpose, functions to couple or combine multiple optical data signals. Hence the design of POF couple-based WDM is feasible. A low-cost solution will be presented for POF-WDM system implementation.

Multiplexing the wavelength division has several advantages over the other approaches proposed to enhance a link's capacity:

1. Deals with appliances of low speed [12]

2. Works for existing single mode cable [12]

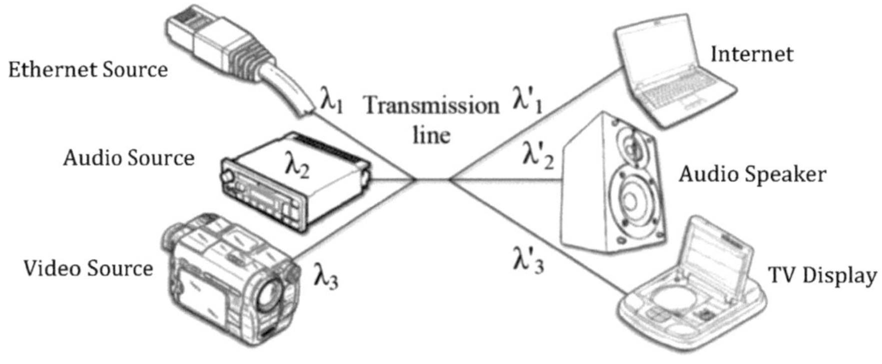


**Figure 3.**
*Proposed design for WDM-POF network in star topology able to transmit three different signals: Ethernet ($\lambda_1$), audio system ($\lambda_1$) and CCTV ($\lambda_1$) at the same time through one fiber.*

3. *Is transparent*: does not rely on the protocol to be transmitted [2, 12]

4. If consumers need it, it is simple for network providers to add additional capacity in a few days. It offers WDM businesses an economic benefit. Part of a fiber can be leased to a consumer who gets quick access to networks without link to others. On the other side, the telecom company still has a separate part of the network for other clients [12, 13]

5. *Is scalable*: a new channel can easily be added to existing channels instead of moving to a new technology. Companies only have to pay for the bandwidth they actually need [12, 14].

In the form of an effective transmission medium, a novel fused POF coupler is fabricated to split and recombine a range of wavelengths that represent different signals. Three separate wavelengths are used for the propagation of three specific networks sources: network link, infotainment network and video transmission system. With a 665 nm wavelength red LED capable of downloading and uploading information via Ethernet cable, while a 520 nm green LED could transmit a video image created from DVD player, and a 470 nm blue LED is an video transmission network for CCTV system within the building.

The coupler and the receiver ends will be filtered with special interference so that the whole WDM system can select a single signal as needed. These solutions for interference filters are known for the visible spectrum and can be used not only in the infrared.

To removed unnecessary signals and select the wavelength of the system as desired for the filter design. Other parameters will be observed, e.g., the effect of the filter position and the efficiency of the WDM-POF system, including optical output power, power loss, optical noise ratio and sensor crosstalk.

The color filters are made up of two types of plastic, almost the same as POF material About 65% of the line is made of polycarbonate plastic co-extruded. The remainder of the line is deep dyed polyester [15, 16]. By subtracting those color wavelengths, filters generate light. A red filter, therefore, absorbs blue and green so that only the red wavelengths can pass. The method is non-additive subtractive, so a full spectrum must be emitted by the light source. The swatch book provides detailed information about each filter's spectral energy curve. The curve defines the color wavelengths that each filter transmits. Supergel 342, for instance, transmits

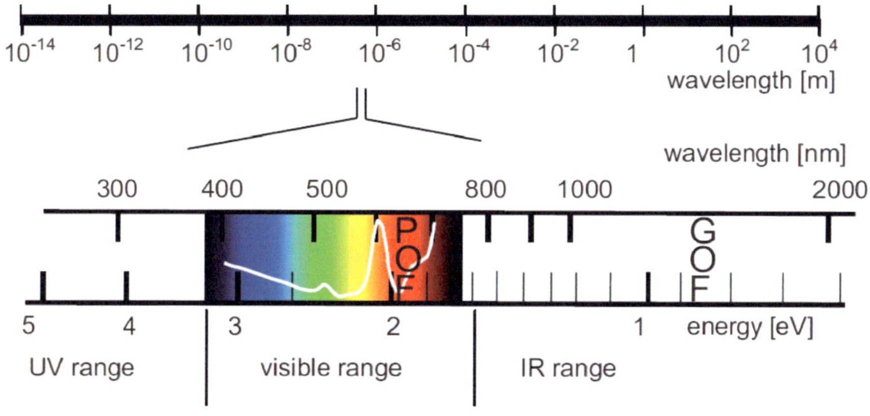


**Figure 4.**
*Attenuation behavior of a POF in the area of the visible spectrum. (source: Gupta [2]).*

about 40% of the blue and violet energy in the spectrum and 70% of the orange and red energy. In the yellow and green range, it absorbs all energy [15, 16].

In this analysis, several colors of red, blue and green filters are evaluated and selected for optimum experiment performance. By reading the curves of the spectral energy distribution (SED), the way the filter colors are selected. In the infrared range above 700 nm, filters that transmit high levels at 700 nm can also transmit high levels. The visible red light, for example, has a wavelength of around 665 nm. Red's filter color is chosen by which film offers the highest percentage of transmission and minimal loss. The same goes with choosing green and blue color filters. Eleven different colors were chosen for each red, green and blue filter in this experiment The goal is to observe which one of the options is better, showing maximum transmission and minimal losses.

The demultiplexer is produced by attaching the multimode POF at one end of the fiber with a connector. A small piece of color films are cut out and prepared for installation on the socket The glue used in this fabrication is epoxy resin, which consists of resin and hardener as mentioned before. When both the resin and the hardener are mixed a strong adhesive is produced that holds the components to be rigidly attached together. Using resin, the small piece of color film is then applied to the edge of the socket after polishing the end of the fiber connected to the socket. Instead, after applying the film on the resin to be applied to the socket, the part is then held tightly together for about 2 minutes to ensure that there is no gap between them and that a strong bond produced. This part has to be done carefully since it is important to avoid the epoxy resin coating the surface of the fiber as much as possible so that any power losses can be avoided when measuring. Nevertheless, since the socket edge is quite thin and sharp, it is not possible to avoid the spread of the epoxy resin to the fiber surface 100%.

After the adhesive is sealed, the POF is placed in a secure location so that the adhesive is not contaminated and is left to dry up. It usually takes nearly a day to completely set the resin. Upon curing of the epoxy, the film that attached to the end of the connector is cut in circle according to the shape of the socket's end. After the production process is carried out, the injection loss and energy consumption for each POF is measured and recorded. The power meter is the device used for reading. Most samples were produced in this test to get the best results and see which of the color filters showing the most transmission and providing the least losses. The POF's length is set at 3 meters long.

As shown in **Figure 5**, The test bed has been set up for 1 × 3 WDM-POF networks to calculate the performance of the handmade coupler-filter combination for the entire system. Red, green and blue LED transmitters inject each of the red filters and readings are taken accordingly. The same is done for the filters blue and red. It is found that measurements should be visible on the meter when calculation is performed with the power meter, otherwise the samples cannot be used to evaluate the characterization. Next, the POF with film must be attached to another short fiber using connector to calculate and obtain the readings. Then the other end of the short fiber to the power meter socket will be connected. The power meter comparison for red LED is set to-10.7 dBm, as is the case for blue and green LED. The other end of the POF (that without film) is attached to the transmitter before the readings can be taken. The LED is then inserted by the fiber and the injection loss rate and energy consumption is measured for each sample accordingly.

The transmitters carry a certain amount of information or data when each transmitter carries signals of different wavelengths defined by the LED. When filtering any other wavelength, the specially tailored color films are used. This requires only a photon to travel through the image and so transmits the data sent to the receiver. In **Figure 6** it is simple to understand filtering and signal coupling operation and **Figure 7** for the experiment of 1 × 2 POF-WDM system through a

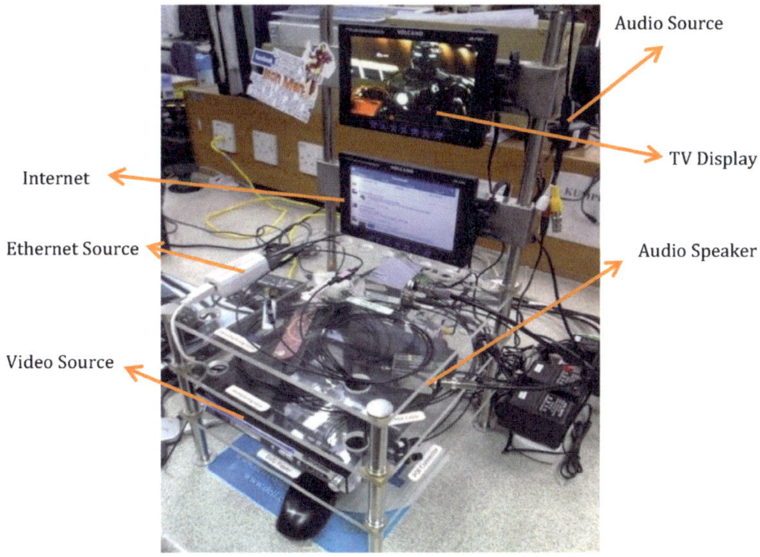


**Figure 5.**
*Test bed for 3-channels WDM-POF system to transmit three different (ethernet, CCTV and DVD player) signals.*

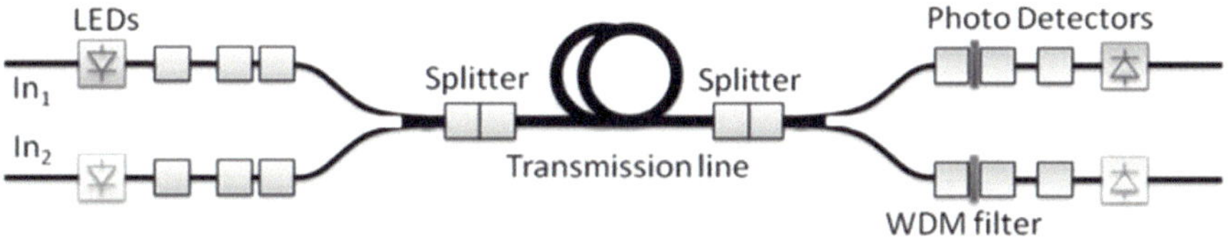


**Figure 6.**
*WDM-POF system design using 1 × 2 handmade coupler and filters.*

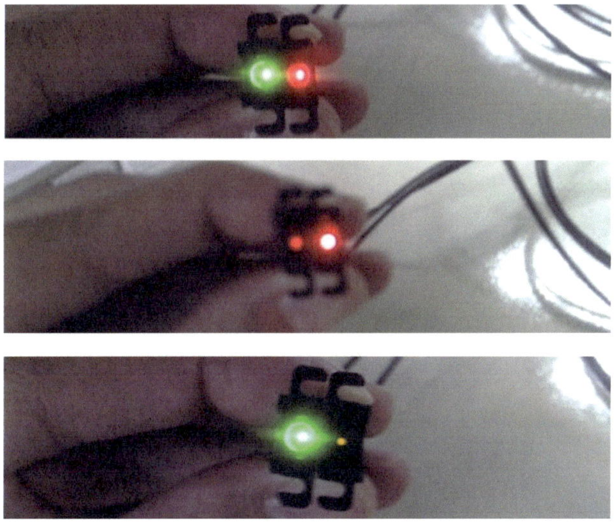

**Figure 7.**
*1 × 2 demultiplexer is used to split the signal to different frequency (color). The multiplexed signal is separated according to the application (e.g. data & video signal) respectively.*

combination of red and green filter. **Figure 7** explained the activity when one of the source deactivated, the crosstalk between both fibers occurred affect the efficiency both transmission, red and green LED.

## 5. Results

The 1 × 3 coupler design plays an important role in combining three optical signals by the fused taper twisted component (see **Figure 2**) in which all three POFs fused and combined as so-called single POF. The fused tapered POFs should be manufactured and all bundled fibers fully fused. Otherwise, it would likely not be possible to pass on the signal led to a failure when combining the single signal numbers. [10, 17, 18].

Either during manufacturing processes or during characterization test stages the error may occur. Controlled heat anomalies during the fusion process become one of the major problems because it makes the core structure of POF more responsive to the heating process. When impaired, it becomes impossible or even difficult to let a light pass through the core To avoid micro-scaled cracks on core it's important to stop twisting and tightening POF. For that cause, if indirect heating is done through fiber, we use the metal tube to reduce the damage to the system.

The excessive deformation in the fused fiber bundle was reduced by indirect heating. It makes the fabrication of the fused-tapered fiber simpler and more effective. In particular, the constant processing capability ensures that fabrication time and productivity is minimized. This approach would significantly reduce the cost of making the coupler [6].

Bidirectional optical loss calculation was performed to investigate exactly the value of the energy intensity for each fused bundle POF output, whereby the red LED was inserted on both sides of the Fused Bundle independently through each POF input.

For both directions (left and right), the average optical loss for the fused POF bundle was measured and analytically compared. The analysis can be seen as shown in the **Figure 8**.

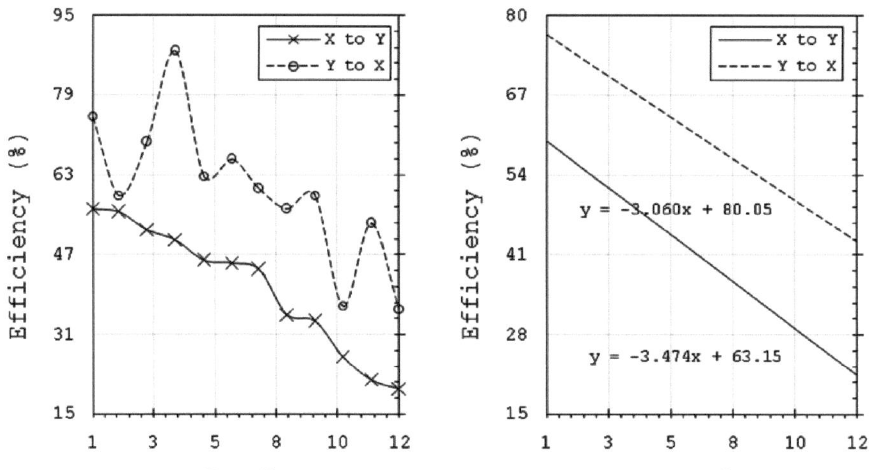


**Figure 8.**
*Efficiency for fused bundle fibers from best to worst sample in both directions and the linear function of the 1 × N.*

The finding above indicates that the optical loss in different directions for the fused bundle was not comparable. The fused POF bundle has an evaluation in the right direction of reduced optical loss. Therefore, it has been chosen as a POF coupler from the right side of the POF bundle because it can integrate multiple optical signals and produce an optical signal with decreased attenuation and greater efficiency than the other. Nevertheless, optical losses for the fused bundle are caused primarily by physical modification of the POF, especially of the Fused Taper.

The improvement in the initial POF diameter resulted in a significant change in optical properties with the numerical aperture and maximum acceptance angle. All these improvements are based on the principle of spoil light propagation; there are more refracted beams of light and they are scattered out beyond the atmosphere [19].

In terms of market value, correlation of handmade and commercial couplers was observed. The total cost of a 1 × 4 handmade POF coupler is less than 3 USD but not less than 250 USD for a commercial coupler available on the market. Currently, many devices are in use to couple a signal, such as a low-cost plastic optical fiber coupler from 1 × 2 acrylic-based [11]. Nevertheless, since the manufacturing processes were very difficult and costly, the handmade POF coupler can be considered a potential solution to this issue.

This research categorizes the optical loss as extrinsic loss because of the physical change of POF, LED projection to POF and the core-to-core connection end [3, 16]. The physical change in POFs caused by the fabrication process is observed to decrease by POFs in diameter to 1 mm and POFs have eventually fused into tapered form. Optical degradation may be caused by the direct LED projection to the POF surface when analysis takes place. In addition, the connection between the fused tapered POF and the POF cable can cause optical loss [20].

The other factor which is critical in transmitting two separate wavelength signals on transmitters is the filter between the coupler and the receiver point. Two separate LEDs were used in this research, related to video signals (CCTV and DVD player); blue LEDs (470 nm) transmit DVD player images through fiber and green LEDs (570 nm) for CCTV, so that the high-quality video signals can be viewed on a monitor screen.

The filter itself has also been tested for its effectiveness. The correlation comes from both green and red LED output during the propagation of a different signal to be divided by a POF connector and the optical energy meter was placed directly before the port of transmission of the transmitter, as shown in **Figure 9**.

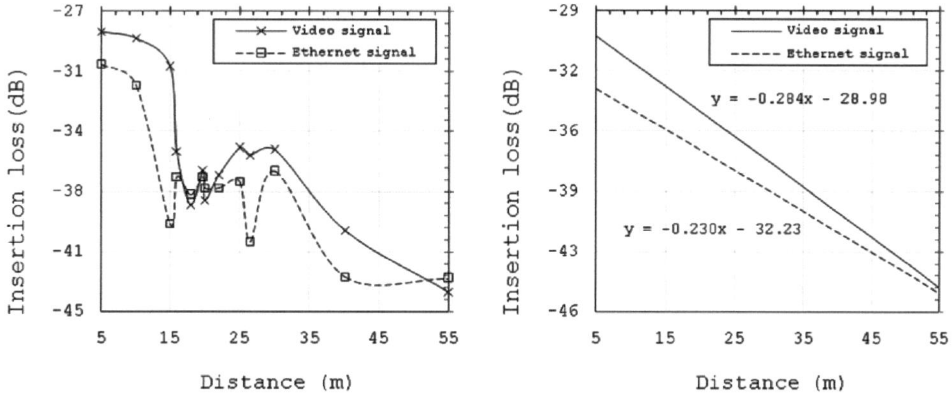


**Figure 9.**
*(a) Power loss comparison between blue and green LED and (b) in linear function, green LED represent the video quality of the CCTV while blue LED represent the DVD player image quality.*

The blue LED shows a higher loss as compared to the red LED in **Figure 9**. DVD images are highly sensitive to varying distances, the greater the distortion, the more the distortion resulted in the screen, the less performance of fiber data transfer.

The variations between the two signal rates were 3 dB and the performance of the video transmission system on the low-cost WDM-POF platform was increased. **Figure 10** demonstrates the video quality using the WDM-POF process.

Comparison for the optical line either using the filter or not, has been analyzed. The insertion loss of the cable with or without red filter is visualized in **Figure 11**, also with it logarithm and linear function of the data.

From the result, the insertion loss measured by the power meter is showing small loss rates when a film is attached to its socket when all the components are configured and red LED is injected. This also applies when injecting blue and green LEDs. We are taking red film A (filer labeled #4690) as the main filter for characterizing the same film using different sources and inject it with all three LEDs, red, green and blue. The results show a small increase in losses, compared to initial losses before the film is attached to the fiber, in the red-injected filter with a Red-outlet-transmitter. When the film is attached to the fiber it is the same for the power output. As **Figure 12** shows, an injection loss of 1 dB was reported right after the resin was inserted into the connector.

An increase of 5.3 μw of power output is observed. This is expected since the used of epoxy resin and the transmission limitation of the film gives the obtained data. However, different results are observed when blue and green LEDs are injected to the red filter. Small decrease of losses is observed when the fiber is attached with the film compared to before the fiber is attached with the film. The utilization of epoxy resin is ruled over by the higher transmission of green and blue transmitter through the particular red film. Above case happens when

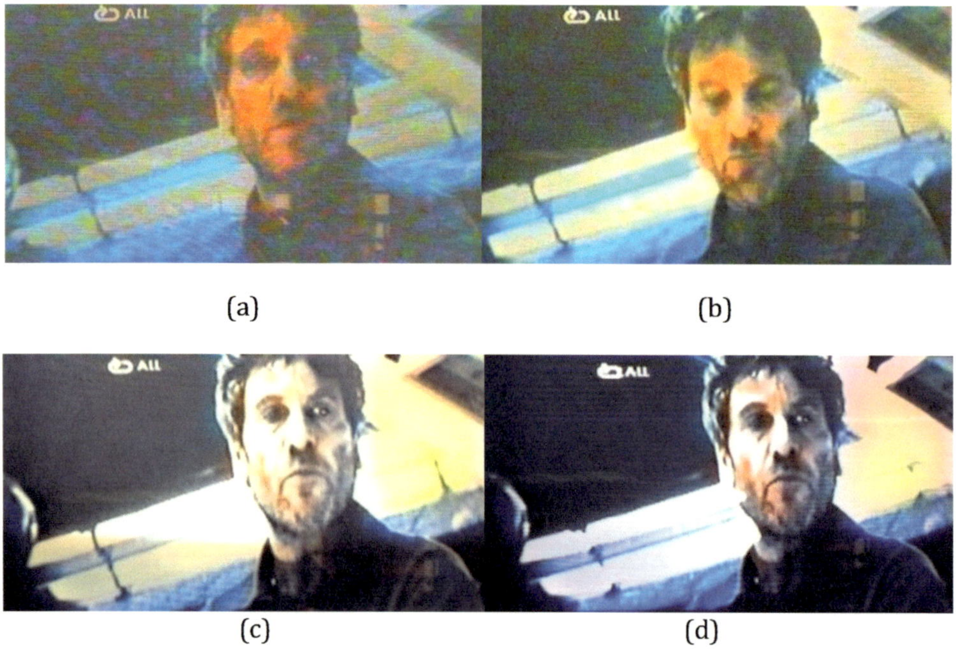

**Figure 10.**
*Video quality of WDM-POF system of (a) 50 m, (b) 30 m, (c) 20 m and (d) 10 m of optical transmission line.*

the transmission percentage of the particular red film also shade or covers some percentages of green and blue wavelength region.

For characterization of same source injected through different filters, red LED is taken as the primary source. From the result, it is observed that sample 7 shows least losses and decrease of efficiency, while sample 3 shows the opposite. Since sample 3 being the one among the darkest film color meaning that only small or narrow transmission percentage of red LED or transmitter is allowed

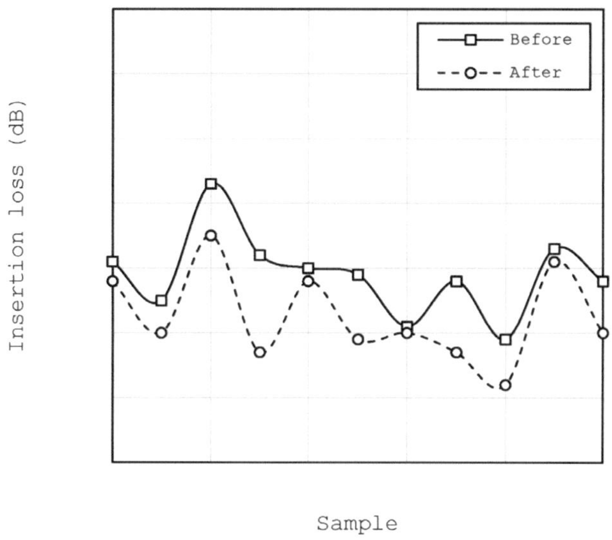


**Figure 11.**
*Effect on resin for demultiplexer filters approximately 1 dB insertion loss occurred on the measurement between before and after connector glued by epoxy.*

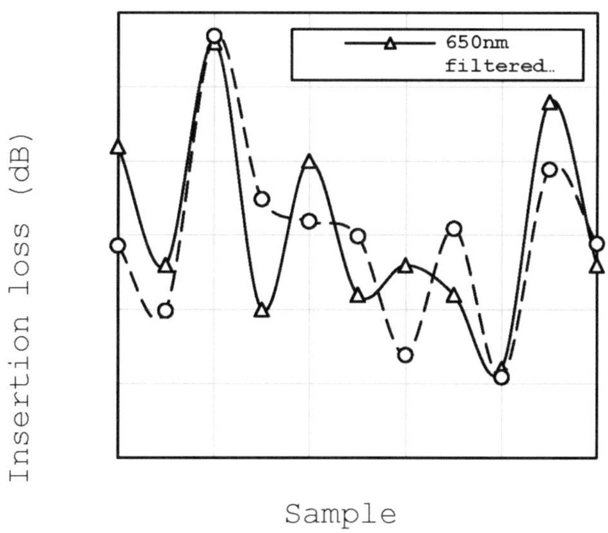


**Figure 12.**
*Comparison between result from experimental and theory measured from red filter signal injected by red LED with 665 nm wavelength.*

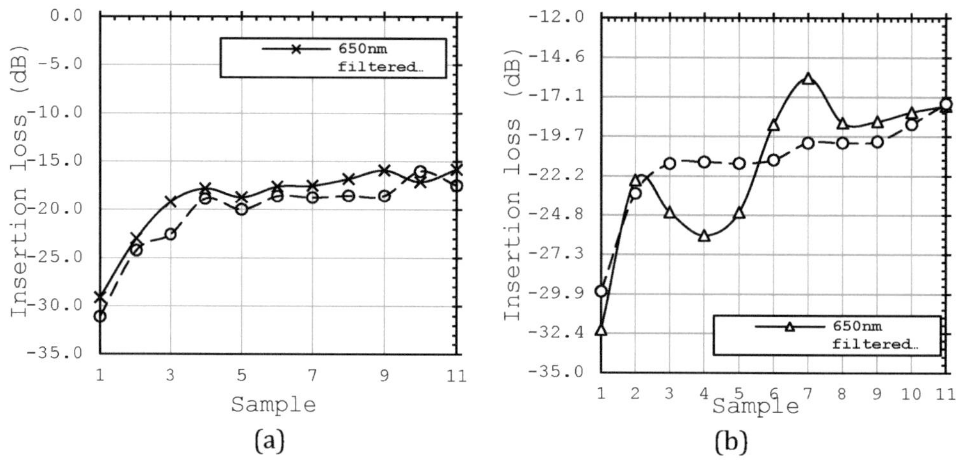


**Figure 13.**
*Comparison between result from experimental and theory measured from (a) green and (b) blue filter signal injected by red LED with 665 nm wavelength.*

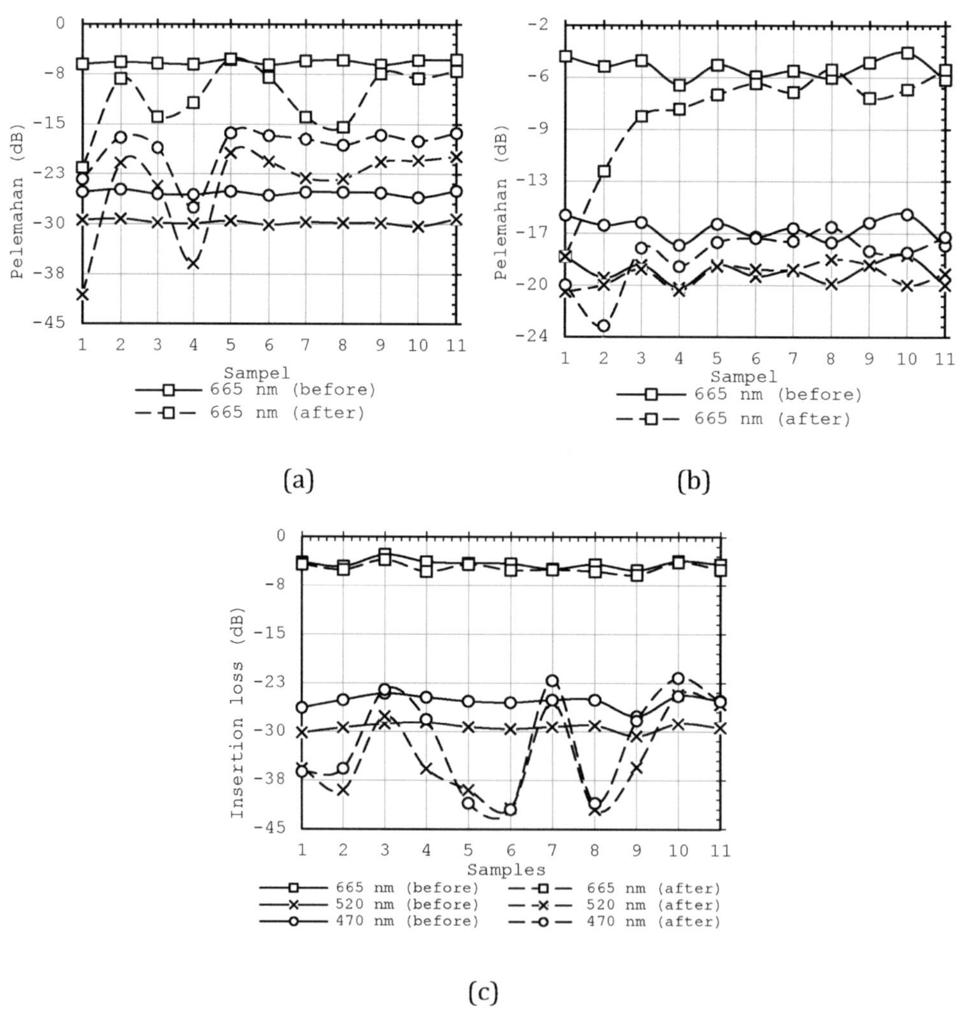


**Figure 14.**
*Effect on combination of working function between optical sources and filters, whereas all three filters (red, green and blue) were injected by light source (a) $\lambda_1$ = 470 nm, (b) $\lambda_2$ = 520 nm and (c) $\lambda_3$ = 665 nm.*

to get through. The utilization of epoxy resin may also contributes to the deficiency of power output (efficiency) since the losses increase a lot for sample 3 apart from the reason it being a dark film with small percentage of transmission. On the other hand, the effect of efficiency of sample 7 is small because the larger percentage of transmission for red LED. According to the ROSCO SED Swatch Report. The predicted outcome can now be correlated with the expected signal from the color filter hypothesis, from the sample setup and the actual signal. The figure shows that on average less than 1 dB is the difference between experiments and theory.

The green and blue films technically are filtered out or block red LEDs due to a different wavelength spectrum of red transmitters that are released by various color films of different wavelengths, such as green and blue films. The red LED data analysis shown through green filters reveals that sample 1 blocks most transmission in areas with approximately 35 dB lower efficiency and the lowest loss among all samples (see **Figure 13a**). The same applies to blue filters, which mostly block the red LED transmission (see **Figure 13b**). The lighter color films (green and blue) the smaller the output, in this case red LED, with various wavelength sources. The lightweight color film permits the transmitting of red LED.

Contrast the red filters with green filters and the blue ones, block some red LEDs as it is clear that the film can only be crossed by red wavelengths ($\tau$ = 665 nm). The film blocks any other propagation not within the spectrum of the wavelength. This principle is the main idea for the demultiplexer model.

Furthermore, combination of working function between LED sources and color filter plays an important role in WDM-POF system. Some samples each attached with three different color filter (blue, green and red filter) has been injected with three different sources (blue, green and red LED) and the graph of each insertion loss has been plotted in **Figure 14** below.

See **Figure 14** above, all data have more variations directly after light hit the filter via fiber as blue filter was inserted by all three optical sources blue green and red color. Likewise, all light sources will fluctuate the data transmission capacity except the blue LED. This fluctuation was caused by the SED percentage of each filter and by the intensity of each light source. The greater light source strength was transmitted by the more fluctuating graph and the less significant of SED percentage deviation was the less fluctuating effect.

## 6. Conclusions

To summarize, the idea of a single channel or wavelength for POF is extended before the WDM definition is introduced, which results in a restriction of bandwidth. By increasing the bandwidth of the Ecofriendly WDM-POF solves this problem. The theory of WDM indicates efficiency, which in short-distance communication has become the alternative. An optical division was made using multimode SI-POF type with 1 mm core size based on POF technology. The coupler was developed through fabrication and characterization stages. A technology was also employed to develop a short-haul communication demultiplexer based on optical polymer fiber.

This experiment shows multiple signals received via a single fiber of different wavelengths. This model was based on the principle of multiplexer and demultiplexer. The system uses just three wavelengths: blue ($\lambda_1$ = 430 nm), green ($\lambda_2$ = 570 nm), and red ($\lambda_3$ = 665 nm) to transmit the transmission components as well as demultiplexer filters. The red, green and blue light source are mixed by multiplexer and separated by demultiplexer.

Filters play a key role in giving Ecofriendly WDM-POF device a greater insertion loss, but because of the filter's color band gap, the internet speed is always constant and the video image resolution is quite good, the performance of a range of output ports has not been badly damaged. Several parameters have been noted, such as optical output power and energy losses on the devices and not to mention the impact of filter placement and efficacy of the handmade 1 × N Ecofriendly WDM-POF coupler.

For characterizing analysis the power level of the demultiplexer has been studied in red LEDs with a 665 nm wavelength inserted into various color filters. Analysis shows that filters of the same wavelength as the transmitter retain performance when other wavelength ranges are either filtered out or blocked This main concept is fully utilized for the designing of demultiplexer for short-haul applications. Final analysis indicates that filter efficiency can exceed 70%. Performance enhancement can be accomplished through practical means. Although the integration device exhibits very high transmitting attenuation, it has been tested for the sending of audio, DVD player and CCTV images data, using this method of handmade Ecofriendly WDM-POF coupler.

The results show that Ecofriendly WDM-POF coupler can be used as a cost-effective wavelength multiplexer, as it can combine different wavelengths with main advantages that are low optical loss and inexpensive. An extensive experiment to enhance the homogeneity of this model was proposed. Nevertheless, fusion methodology with certain drawbacks has no uniformity in fabricating Ecofriendly WDM-POF coupler, as POF coupler with good performance could hardly be manufactured reliably. With experience and practice, this Ecofriendly WDM-POF system can be enhanced. The Ecofriendly WDM-POF system is very preferred because it is not as pricey as other consumer POF coupler. In addition, the production and deployment process is simple, quick and suitable for implementation for short distance communication.

## Acknowledgements

This research has been conducted in Computer& Network Security Laboratory, Universiti Kebangsaan Malaysia (UKM). This project is supported by Ministry of Science, technology and Environment, Government of Malaysia, 01-01-02-SF0493 and Linear Dms Solutions Sdn Bhd, DIP-2018-017. All of the handmade fabrication method of POF coupler, 1 × N handmade™-POF coupler and also the low cost WDM-POF network solution were protected by patent numbered PI2010700001.

## Author details

Hadi Guna*, Mohammad Syuhaimi Ab-Rahman, Norhana Arsad, Roslan Shukor and Sahbudin Shaari
Universiti Kebangsaan Malaysia, Selangor Darul Ehsan, Malaysia

*Address all correspondence to: hadi_guna87@yahoo.com

## References

[1] KK, HP, Photodetectors for LiDAR. Hamamatsu Corporation; 2019. p. 20

[2] Gupta SC. Textbook on Optical Fiber Communication and Its Applications. 3rd ed. New Delhi: PHI Learning Pvt. Ltd; 2018

[3] Janota A, Hrbček J. Slovak ETC system implemented–what next? In: Mikulski J, editor. Transport Systems Telematics. Berlin Heidelberg: Springer; 2011. pp. 30-37

[4] Ericsson Green Power to Bring Mobile Telephony to Billions of People. 2008

[5] Schubert EF et al. Solid-State Lighting- : A Benevolent Technology. Vol. 69. Bristol: ROYAUME-UNI: Institute of Physics; 2006. p. 31

[6] Jeong Y, Bae S, Oh K. All fiber N × N fused tapered plastic optical fiber (POF) power splitters for photodynamic therapy applications. Current Applied Physics. 2009;**9**(4, Supplement 1): e273-e275

[7] Imoto K et al. New biconically tapered fiber star coupler fabricated by indirect heating method. Journal of Lightwave Technology. 1987;**5**(5):694-699

[8] Kagami, M., Visible Optical Fiber Communication, in Special Issue: Visible Optical Fiber Communication. Japan: R&D Review of Toyota CRDL; 2005;**40**(2):1-6

[9] Ab-Rahman MS, Guna H, Harun MH. 1xN self-made polymer optical fiber based splitter for POF-650nm-LED based application. In: 2009 International Conference on Electrical Engineering and Informatics. Selangor, Malaysia; 2009

[10] Ab-Rahman MS et al. Fabrikasi dan Pencirian Pencerai Optik 1×12 Buatan Tangan Berasaskan Gentian Optik Polimer Diperbuat daripada Polimetil Metakrilat. Sains Malaysiana. 2010;**39**(3):459-466

[11] Ehsan AA, Shaari S, Ab-Rahman MS. Low cost 1x2 acrylic-based plastic optical fiber coupler with hollow taper waveguide. In: 25TH of Progress in Electromagnetics Research Symposium. Beijing, China: Progress in Electromagnetics Research Symposium, PIERS; 2009

[12] Gumaste A, Antony T. DWDM Network Designs and Engineering Solutions. USA: Cisco Press; 2003

[13] Bischoff D. Wavelength Multiplexing: WDM and DWDM Systems. Essay. 2009. pp. 1-27

[14] Gupta P, Khurana H. Public entrepreneurship: A dynamic strength for budding green technology. In: Proceedings of the 4th National Conference; INDIACom-2010. New Delhi; 2010

[15] Rosco, Roscolux, Harbor View Avenue. Stamford: Rosco Laboratories; 2003

[16] Rosco. In: Rosco, editor. ROSCOLUX Color Filter. Stamford: Rosco: Harbor View Avenue; 2003. pp. 1-2

[17] Ab-Rahman MS et al. In: Yasin M, Harun SW, Arof H, editors. Integration of Eco-Friendly POF Based Splitter and Optical Filter for Low-Cost WDM Network Solutions. Optical Fiber Communications and Devices. London: InTech; February 2012. p. 380

[18] Ab-Rahman MS et al. A novel star topology POF-WDM system. In: Business, Engineering and Industrial Applications (ISBEIA), 2011 IEEE Symposium on 2011; 2011

[19] Held G. Fiber-Optic and Satellite Communications in Understanding Data Communications. USA: New Riders Publishing; 2010

[20] Appajaiah A, Kretzschmar HJ, Daum W. Aging behavior of polymer optical fibers: Degradation characterization by FTIR. Journal of Applied Polymer Science. 2007;**103**(2):860-870

Chapter 4

# Study on Designing and Manufacturing a Radio-Frequency Generator Used in Drying Technology and Efficiency of a Radio Frequency-Assisted Heat Pump Dryer in Drying of *Ganoderma lucidum*

*Nguyen Hay, Le Anh Duc and Pham Van Kien*


## Abstract

A radio-frequency (RF) generator applied in drying technology was designed and manufactured for drying *Ganoderma lucidum*. The drying experiments were conducted by drying method of RF-assisted heat pump in order to inspect the operating parameters of the RF generator and investigate the effects of the input drying parameters on drying rate in the RF-assisted heat pump drying of *Ganoderma lucidum*. The results have shown that the RF generator achieved the required operating parameters as design such as RF power of 3 kW and operating frequency of 27 MHz. In RF-assisted heat pump drying, increase in RF power and drying air temperature increases the drying rate. Meanwhile, drying air velocity does not significantly affect the drying rate. At RF power of 1.95 kW, the drying time reduces by 9, 17, and 33% in comparison with RF power of 1.3, 0.65, and 0 kW (heat pump drying). At drying air temperature of 50°C, the drying time reduces by 10% and 21% in comparison with drying air temperature of 40 and 45°C. Besides, increasing RF power retains the higher content of polysaccharide in *Ganoderma lucidum*, and the *Ganoderma lucidum* samples retain the color better after drying.


**Keywords:** *Ganoderma lucidum*, heat pump, radio frequency, drying temperature, drying time

## 1. Introduction

Drying is a common and effective preservation technique that reduces moisture content of material to lower levels required. Therefore, drying can minimize the spoilage of various microbes in material and the physical, chemical, and biochemical changes within the drying products thereby increasing overall shelf life by considerable periods of time. However, the drying process will affect the quality of the product such as nutritional standards, sensory standards, and physical and chemical standards. Therefore, the drying method and drying parameters should be considered to find a suitable drying method with optimum drying condition to retain a high quality of drying products, especially in food technology, agricultural products, and medicinal products.

*Ganoderma lucidum* is a medicinal product that contains various bioactive ingredients. Polysaccharide is a main bioactive ingredient in *Ganoderma lucidum* which has been found to be medically active in several therapeutic effects such as antitumor, anti-inflammatory, antiviral, anticancer, and anti-HIV [1]. However, polysaccharide and other bioactive ingredients in *Ganoderma lucidum* are heat sensitive and the high drying temperature tends to cause higher loss of active ingredients in dehydrated *Ganoderma lucidum*. Therefore, in drying *Ganoderma lucidum*, drying method as well as drying parameters should be considered carefully.

RF technology has shown some unique advantages in drying technology. RF heating is a volumetric heating method, which provides fast and deeper heat generation within material that increases heating rate and shortens drying time significantly. RF heating mechanism is described in **Figure 1**. In which, the RF generator creates an alternating electric field between two electrodes. The material is placed between the electrodes. The wet molecules within material continuously reorient themselves to face opposite poles of the alternating electric field. The friction resulting from the rotational movement of the molecules and the space-charge displacement causes the material to rapidly heat throughout its mass.

There were numerous studies of RF drying technology in which RF is combined with other drying methods as convection drying using hot air and freeze-drying for drying food and agricultural products [2–9]. The results show that heat generation within the whole volume of drying material that supports the heat transfer and moisture diffusion process to take place faster shortens the drying time and the temperature, and moisture distribution within material becomes more uniform. The drying products still retain their characteristic color and taste.

In heat pump drying with circulating drying air, drying air after being blown through the heat pump has the specific temperature, velocity, and humidity. Drying air will be blown into the drying chamber, and the drying process is performed here. In heat pump drying, the drying air temperature is at low level. So, the drying products can retain a high content of bioactive ingredients and their characteristic color and taste.

The drying technology using RF and heat pump drying has been found to be suitable for drying medicinal products. The objectives of this study are (1) to design

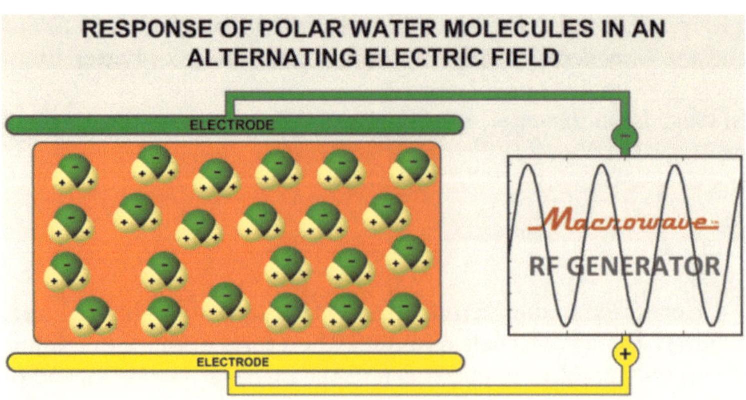


**Figure 1.**
*RF heating mechanism.*

and manufacture a RF generator applied in drying technology for drying *Ganoderma lucidum*, in which RF-assisted heat pump drying method is applied, and (2) to investigate the effects of the input drying parameters as drying air temperature, drying air velocity and RF power on the drying rate, and the quality of *Ganoderma lucidum* in RF-assisted heat pump drying process.

## 2. Researching object and method

### 2.1 Researching object

A RF generator applied in drying technology is designed and manufactured in order to achieve a required maximum RF power of 3 kW and frequency of 27 MHz.

*Ganoderma lucidum* used for the experiments is red *Ganoderma lucidum* (*Ganoderma boninense*). After being harvested, *Ganoderma lucidum* has a moisture content of 3 (d.b) (i.e., 75% (w.b)), diameter of 12 cm, thickness of 1.5 cm, and glossy red brown color. *Ganoderma lucidum* samples are cleaned with dry tissues. The initial moisture content of the material is determined by a moisture analyzer (see **Table 1**).

### 2.2 Researching methods

#### *2.2.1 Designing and calculating method*

- The required RF power is calculated based on physical and thermal properties of *Ganoderma lucidum* and theory of designing and calculating drying system.

- The circuit diagram and the components of the RF generator are designed and manufactured based on theory of RF heating mechanism, heat exchanger, and oscillator circuit of RF generator.

#### *2.2.2 Manufacture of RF generator method*

The components of RF generator are manufactured in a single unit as designed and installed to complete a RF operator. Some standard components are selected and purchased in the market.

| No | Symbol | Value |
|---|---|---|
| 1 | $G_b$ | 20 kg/batch |
| 2 | $m_{LC}$ | 20 kg |
| 3 | $\omega_i$ | 75% (w.b) |
| 4 | $\omega_f$ | 13% (w.b) |
| 5 | $C_p$ | 3.613 kJ/(kg °C) |
| 6 | r | 3150 kJ/kg |
| 7 | $t_i$ | 30°C |
| 8 | $t_f$ | 45°C |

**Table 1.**
*Physical thermal index of the drying material.*

#### *2.2.3 Measurement method*

The parameters can be measured by specialized measuring instruments directly such as temperature, velocity of drying air, voltage, and electric current. The other parameters are determined by the exchange formulas.

#### *2.2.4 Method of experiment*

Experiments for investigation of the effects of the input drying parameters on drying rate in the RF-assisted heat pump drying of *Ganoderma lucidum* are conducted at the drying air temperature of 40, 45, and 50°C; drying air velocity of 1.2, 1.6, and 2.0 m/s; and RF power of 0.65, 1.3, and 1.95 kW.

#### *2.2.5 Method of determining moisture content*

The *Ganoderma lucidum* weight measurements are taken regularly after intervals of 20 minutes by an electronic scale digital balance (see **Table 1**). Each experiment is conducted until the drying material achieves the moisture content of 0.15 (d.b) (i.e., 13% (w.b)) and completed in triplicates.

The color of the drying products is measured by a colorimeter (see **Table 1**). The colorimeter displays three reflected light intensities corresponding to the lab color values. The total change in color of the drying *Ganoderma lucidum* sample with reference to the original sample is calculated as

$$\Delta E^* = \sqrt{(L_0 - L^*)^2 + (a_0 - a^*)^2 + (b_0 - b^*)^2} \tag{1}$$

The parameters in Eq. (1) are described in detail in part of 3.4.2 c. (3.4.2 c. Color of drying material).

Polysaccharide content of *Ganoderma lucidum* is determined by high-performance liquid chromatography (HPLC) method.

Statistical parameters such as mean and standard deviation are used to solve the experiment data. Examining the differences of the statistical data is conducted by means of least significant difference (LSD).

## 3. Results and discussions

### 3.1 The RF power of RF generator

The heat required for drying process was calculated based on the theory of calculating and designing drying system [10].

Physical and thermal property index of the drying material (*Ganoderma lucidum*) is given in **Table 2**.

#### *3.1.1 The heat required for heating the material*

The heat required for heating the material in drying process is the heat of heating the drying material until the material achieves the required temperature. The required temperature of 45°C is chosen for calculation:

$$q_1 = m_{LC}.C_p.(t_f - t_i) = 20 \times 3.613 \times (45 - 30) = 1084\,kJ \tag{2}$$

The predictive time period required for *Ganoderma lucidum* to get the temperature of 45°C is 35 minutes. The heat required is calculated as

$$Q_1 = \frac{q_1}{\tau_1} = \frac{1084}{35 \times 60} = 0.516 kW \tag{3}$$

### *3.1.2 The heat required for vaporizing water in the material*

In the drying process, an amount of heat must be supplied to vaporize the water within drying material at specific drying temperature in order that the material achieves the required final moisture. The heat required depends on the mass of vaporized water in the material and latent heat of drying material.

The mass of vaporized water in the material (kg) is

$$G_W = \frac{m_{LC}.(\omega_i - \omega_f)}{1 - \omega_f} = \frac{20.(0.75 - 0.13)}{1 - 0.13} = 14.25\ kg \tag{4}$$

So,

$$q_2 = G_W.r = 14.25 \times 3150 = 44896\ kJ \tag{5}$$

The initial moisture content of *Ganoderma lucidum* is 75%. The predictive time period required for *Ganoderma lucidum* to get the final moisture content of 13% is 7 hours. The heat required is calculated as

$$Q_2 = \frac{q_2}{\tau_2} = \frac{44896}{7 \times 3600} = 1.782\ kW \tag{6}$$

| No | Name | Description |
|---|---|---|
| 1 | Colorimeter | Type: Minolta CR-200 |
| 2 | Frequency measurement instrument | Type: Acoustimeter CAT #A139<br>Max frequency: 70 ± 0.01 MHz |
| 3 | High-voltage voltmeter | Type: Voltmeter-MDP-50 K<br>Voltage range resolution: 0.5–10 kVAC ±5% |
| 4 | Amperemeter | Type: Amperemeter-C.A401<br>Ampere range resolution: 0.1–10 A ± 1% |
| 5 | Thermal sensor | Type: AYN-MF59-104F-3950FB-1000<br>Measurement ranges: −60–300°C ± 0.05°C |
| 6 | Moisture analyzer | Type: DBS 60–3 model<br>Maximum capacity: 60 g ± 0.01%<br>Temperature range: 50–200°C<br>Temperature increments: 1°C<br>Repeatability (sd) with 2 g sample: 0.15%<br>Moisture value predicted: 0–100% |
| 7 | Electronic scale digital balance | Type: DS-2002-N<br>Max weighing capacity of 2000 ± 0.001 grams |

**Table 2.**
*Parameter index of the measurements.*

### *3.1.3 The heat loss for heating the drying tray*

In the drying process, the drying material is placed on a drying tray which is normally a plastic mesh grid. So, there must be an amount of heat loss for heating the drying tray until the drying tray gets the drying air temperature:

$$q_3 = m_{tray}.C_{PVC}.(t_f - t_i) = 25.05\ kJ \tag{7}$$

in which $m_{tray}$ is the mass of the tray, $m_{tray}$ = 1 kg, and $C_{P\_plastic}$ is the specific heat capacity of plastic, $C_{P\_plastic}$ = 1.67 kJ/(kg °C).

The predictive time period required for the drying tray to get the temperature of 45°C is 45 minutes. The heat loss is calculated as

$$Q_3 = \frac{q_3}{\tau_3} = \frac{25.05}{45 \times 60} = 0.009\ kW \tag{8}$$

### *3.1.4 Heat loss through pipes*

In the drying process, the drying air flows inside a pipe system, and the outside wall of the pipe is in contact with environment. So, the heat loss through pipes should be considered, and it depends on the pipe material, size, length of the pipe, and drying temperature. The pipe is normally made of PVC plastic.

The length of pipe from the pump to the drying chamber is 1.5 m, so the surface area of the pipe is

$$S_0 = \pi.d_2.l = 3.14 \times 0.2 \times 1.5 = 0.942\, m^2 \tag{9}$$

So, the heat loss through pipes is calculated as

$$Q_4 = S_o.q_4 = S_o.\lambda_{PVC}.\left(t_f^v - t_i^v\right).\frac{2.\pi}{\ln\frac{d}{d_1}} = 0.374\, kW \tag{10}$$

in which $t_i^v$ (30°C), $t_f^v$ (45°C), $d_1$ (0.193 m), $d_2$ (0.2 m), and $\lambda_{PVC}$($\lambda_{PVC}$ = 0.15 W/(m °C)) are temperature of the outside wall and inside wall, internal diameter, external diameter of the pipe, and thermal conductivity of PVC plastic.

### *3.1.5 Heat loss for heating the drying chamber*

The drying process is performed in a drying chamber that is also heated up to the drying temperature. So, the heat loss for heating the drying chamber should be considered, and it depends on the material and mass of the chamber and drying temperature. In drying process of food and agricultural products, the drying chamber is normally made of a galvanized steel for food hygiene:

$$q_5 = m_{ch}.C_{steel}.\left(t_f^{ch} - t_i^{ch}\right) = 30 \times 0.49 \times (45 - 30) = 220.5\, kJ \tag{11}$$

The predictive time period required for the drying chamber to get the temperature of 45°C is 25 minutes. The heat loss is calculated as

$$Q_5 = \frac{q_5}{\tau_5} = \frac{220.5}{25 \times 60} = 0.147\, kW \tag{12}$$

in which $m_{ch}$ (30 kg), $t_1^{ch}$ (30°C), and $t_2^{ch}$ (45°C) are mass of drying chamber, initial temperature, and final temperature of the drying chamber and $C_{p\_steel}$ is specific heat of galvanized steel, $C_{p\_steel}$ = 0.49 kJ/(kg °C).

### *3.1.6 The heat loss through the drying chamber wall*

The inside wall of drying chamber is in contact with drying air, and the outside wall is in contact with the environment. This causes the heat loss through the drying chamber wall in the drying process, and it depends on the material and area of the drying chamber. The area of the drying chamber (F) includes the area of the drying chamber wall ($F_w$) and two tops ($F_t$).

The area of the drying chamber wall is

$$F_w = 2.(l_{ch}.h_{ch} + w_{ch}.h_{ch}) = 2 \times (1.25 \times 0.75 + 1.15 \times 0.75) = 3.6\, m^2 \tag{13}$$

After expanding the top of the drying chamber on computer by AutoCAD software, the surrounding area of a top is$F_t = 1.62\, m^2$.

So, the area of the drying chamber is

$$F = F_w + 2.F_t = 6,847\, m^2 \tag{14}$$

The heat loss is calculated as

$$Q_6 = q_5.F = k\left(t^{ch}_{inside} - t^{ch}_{outside}\right).F = 0.212\; kW \tag{15}$$

in which k is thermal conductivity of galvanized steel and k is 2.06 W/(m.°C). $l_{ch}$, $w_{ch}$, and $h_{ch}$ are the length, the width, and the height of the drying chamber.

### *3.1.7 Radiation heat loss*

The radiation heat loss is calculated as

$$Q_7 = \varepsilon.F.C_o.\left[\left(\frac{T_f}{100}\right)^4 - \left(\frac{T_i}{100}\right)^4\right] = 0.593\, kW \tag{16}$$

in which ε is the radiation ratio of galvanized steel, ε = 0.85, and $C_0$ is the radiation ratio of absolute black object, $C_0 = 5.67W/\left(m^2.K^4\right)$.

Thus, the total heat required for drying process is

$$Q_{total} = Q_1 + Q_2 + Q_3 + Q_4 + Q_5 + Q_6 + Q_7 = 3.632\ \text{kW} \tag{17}$$

In current study, the RF operator will be designed, manufactured, and applied in RF-assisted heat pump drying. So, in the drying process, RF heating has the main function of heating the material, vaporizing water within the material, and heating the drying tray. The other heat losses are supplied by heat pump. Thus, the heat required for RF generator is.

$$Q_{RF} = Q_1 + Q_2 + Q_3 = 2.307\ \text{kW} \tag{18}$$

Therefore, the RF power of RF generator is chosen P = 3 kW.

## 3.2 Circuit diagram of RF generator

The circuit diagram of RF generator was designed based on the theory of RF heating mechanism, heat exchanger, and oscillator circuit of RF generator [11]. The circuit diagram of RF generator is described in **Figure 2**.

### *3.2.1 Power supply unit*

The power supply unit consists of a transformer, a wire supply voltage transformer, and a rectifier.

The transformer has the function of changing three-phase voltage 380 VAC into 6.5 kVAC. This high voltage is converted into DC voltage of 6.5 kVDC by the rectifier and supplied to the oscillation circuit. Besides, the wire supply voltage transformer will change the voltage from 380 VAC to 12.6 VAC to supply the triode tube filament.

### *3.2.2 Oscillation circuit*

The oscillation circuit consists of a high-frequency triode tube and LC oscillation circuits. A high voltage of 6.5 kVDC is applied to the anode of the triode tube after passing through an induction circuit including L1, L2, and C1 that acts as a filter circuit to remove the alternating current components of the supply power.

A high voltage of 12.6 VAC is applied to the filament and grid pin of the triode tube. A 12.6 VAC power is applied to the grid pin of the triode tube through an induction circuit that consists of $L_5$, $L_6$, and $C_4$. The induction circuit controls the voltage of the grid pin to generate the output frequency at 27 MHz.

### *3.2.3 RF emitting circuit*

The RF emitting circuit is a circuit consisting of $L_3$ and $C_3$ in parallel. The RF high-frequency energy at the output of the high-frequency triode tube passes

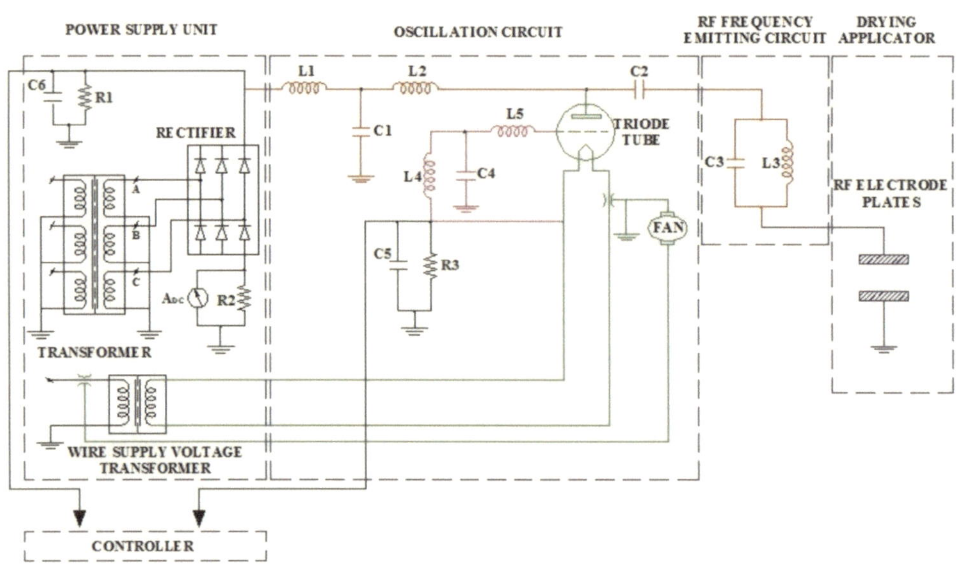


**Figure 2.**
*The circuit diagram of RF generator.*

through the RF emitting circuit, and it is supplied to the electrode plates of the drying applicator.

#### *3.2.4 Drying applicator*

The drying applicator is composed of two parallel electrode plates which are called RF electrodes. The drying material is placed between the electrodes during drying process. The material is heated based on dielectric heating principle.

### 3.3 Fabricating the components of the RF generator

#### *3.3.1 High-frequency triode tube*

The high-frequency triode tube is selected in the market according to the required RF power, and it has the specific specifications as follows:

- Type: Toshiba 7T69RB.
- Voltage applied to filament: 12.6 VAC.
- Frequency: 27 MHz.
- Voltage applied to anode: 6.5 kVDC.
- Output power (maximum): 5 kW.

The output power of 5 kW will be converted to RF electrode plates in drying applicator at 60% efficiency (**Figure 3**).

#### *3.3.2 Power supply unit*

The transformers and the rectifier were manufactured at the workshop with the engineering specifications required. The transformer and rectifier are shown in **Figures 4** and 5.

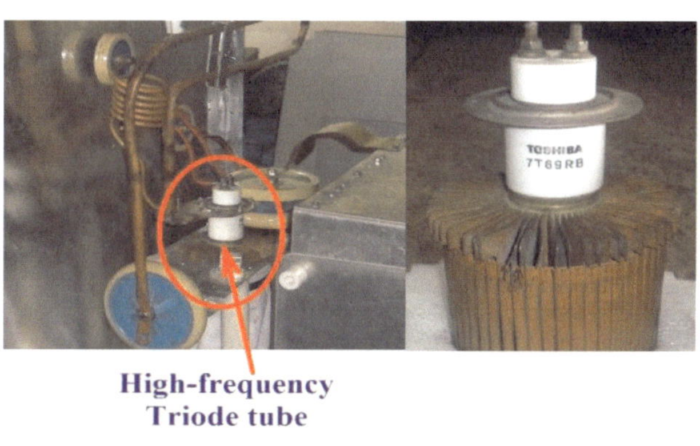


**Figure 3.**
*High-frequency triode tube.*

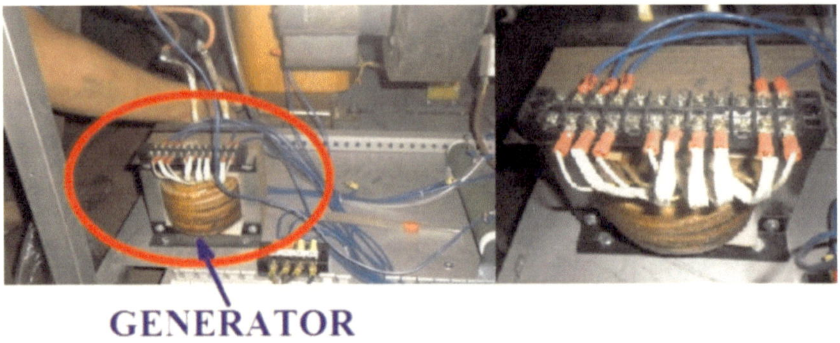


**Figure 4.**
*Transformer.*

**Figure 5.**
*Rectifier.*

### *3.3.3 Oscillation circuit*

The oscillation circuit consists of numbers of capacitors and the inductor coils. The function of the oscillation circuit is amplifying the power and required generating frequency. The capacitors and the inductor coils are the industrial components that can work at the high voltage and high frequency (**Figure 6**).

The oscillation circuit consists of two induction circuits:

1. The $L_1$, $L_2$, and $C_1$ induction circuit works as a filter circuit to remove the alternating current components.

2. The $L_4$, $L_5$, and $C_4$ induction circuit regulates the voltage at the grid pin of the high-frequency triode tube to generate the output RF.

These capacitors and the inductor coils are selected and manufactured according to the standard in Strayfield's handbook for manufacturing RF generator [11], in which the capacitors $C_1$ and $C_4$ are selected in the market, while the inductors $L_1$, $L_2$, $L_4$, and $L_5$ are manufactured at the workshop. Their values are as follows:

$$C_1 = C_2 = 1500 \text{ pF}; C_4 = C_5 = C_6 = 500 \text{ pF}; L_1 = L_2 = L_4 = L_5 = 1.5 \times 10^{-6} H.$$

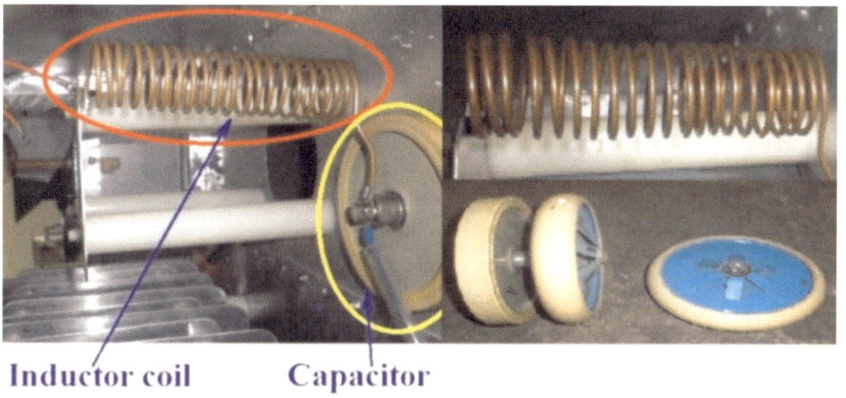


**Figure 6.**
*Inductor coil and capacitor.*

### 3.3.4 RF emitting circuit

The structure of RF emitting circuit is composed of $L_3$ and $C_3$ in parallel that forms an induction circuit. The RF emitting circuit has the function of generating operating frequency of 27 MHz that is the technical requirements. $L_3$ and $C_3$ are manufactured in the workshop according to the technical requirements with the specifications below.

The structure of the capacitor $C_3$ consists of two parallel electrode plates. The capacitance value of capacitor $C_3$ depends on the area of the parallel electrode plates and distance between them.

The electrode plates have an area of $A_{C_3} = 0.45 \times 0.45 = 0.203\, m^2$, and the distance of two electrode plates is $d_{C_3} = 0.1\, m$.

The capacitance value of capacitor $C_3$ is calculated as

$$C_3 = \frac{\varepsilon_0 . A_{C3}}{4\pi . k . d_{C3}} = 1.8 \times 10^{-11}\, F \tag{19}$$

in which $\varepsilon_0\,(\varepsilon_0 \approx 1)$ is dielectric constant of air between two plates and k $(k = 9 \times 10^9 N.m^2/C^2)$ is electrostatic constant.

The function of the $L_3$ and $C_3$ induction circuit is generating operating frequency of 27 MHz. So, the inductance value of $L_3$ is calculated with the parameter f = 27 MHz as follows:

$$L_3 = \frac{1}{(2\pi . f)^2 . C_3} = 1.9x10^{-6}\, H \tag{20}$$

The inductor $L_3$ is manufactured in workshop with its specific specification as follows:

- Inductance value: $L_3 = 1.9x10^{-6}\, H$.
- Material: a copper wire.

- Diameter of wire: 2.5 mm.
- Diameter of wire coil: 40 mm.

*3.3.5 Drying applicator*

The drying applicator consists of two electrode plates which are called RF electrodes. The RF electrodes are fabricated at the workshop. The material used for fabrication of RF electrodes must be a good electric conductive material, and aluminum is chosen. The electrodes have a rectangular surface and dimension of 1200 mm × 1100 mm. They are fixed in drying chamber and connect to the RF emitting circuit through thin copper connectors. The distance between two electrodes is fixed by Teflon plastic bars. The RF electrodes are shown in **Figure** 7.

### 3.4 Drying experiment results

The RF-assisted heat pump dryer used in drying experiment is shown in **Figures 8** and **9**. In the drying process, the drying air is circulated over the evaporator of heat pump. The evaporator cools the drying air further down below the condensation temperature. Below this temperature, the drying air will be dehumidified. Then, the drying air is heated to the desired temperature inside the condenser and blown inside the drying chamber for drying process. In the drying chamber, the drying air will combine with the RF generated by the RF generator to conduct drying process of *Ganoderma lucidum*.

In drying experiment, the mass of *Ganoderma lucidum* selected is 4 kg. Thus, the RF power is adjusted to achieve the value of 0.65, 1.3, and 1.95 kW.

*3.4.1 Result of operating parameters of RF generator*

The RF generator is operated with the maximum RF power of 3 kW to inspect the operating parameters. The operating frequency of RF generator (f) is measured by a frequency measurement instrument, the operating voltage (U) is measured by a high-voltage voltmeter, and the operating current (I) is measured by an amperemeter. The temperature of the material in drying process is measured by a thermal sensor that is connected to a computer through an integrated circuit. The temperature is recorded each 2 minutes.

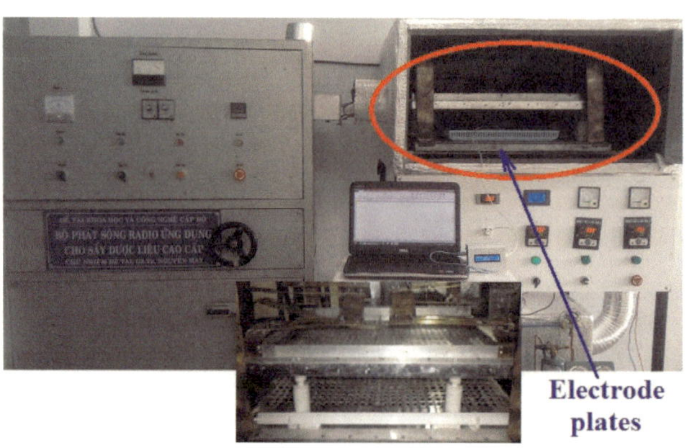


**Figure 7.**
*Electrode plates.*

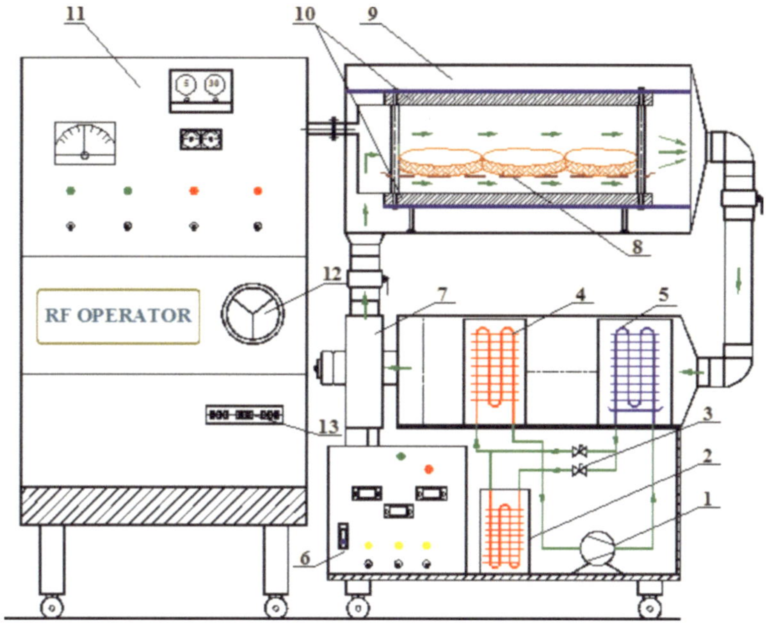


**Figure 8.**
*RF-assisted heat pump dryer model. (1) compressor, (2) sub-condenser, (3) valve, (4) condenser, (5) evaporator, (6) heat pump controller, (7) air fan, (8) drying tray, (9) drying chamber, (10) RF electrodes, (11) RF operating controller, (12) operating current intensity controller, (13) unit of supplying the operating voltage.*

**Figure 9.**
*RF-assisted heat pump dryer.*

The measurement of the operating parameters of RF operator has got the results as follows:

- f = 27 MHz, U = 6.5 kV, I = 0.46 A. So, the power P = U.I = 2.99 kW.

- The material is heated and achieves the required temperature of 45°C in 28 minutes.

The results show that the operation parameters achieve the designing requirement.

The engineering parameters of measurement instruments are described in the **Table 1**.

### *3.4.2 Evaluation of the effect of RF power*

#### *3.4.2.1 Drying time*

The drying curves of RF-assisted heat pump drying process at the drying air temperature of 45°C, drying air velocity of 1.2 m/s, and RF power of 0.65, 1.3, and 1.95 kW is presented graphically in **Figure 10**.

As shown in **Figure 10**, increasing RF power has a significant effect on moisture ratio; the moisture ratio is higher at higher RF power. At RF power of 1.95 kW, the drying time reduces by 9, 17, and 33% in comparison with RF power of 1.3, 0.65, and 0 kW (heat pump drying). It can be explained by RF heating mechanism, in which, increasing the RF power will increase energy absorption inside *Ganoderma lucidum*, which makes water dipole molecules and free ions in *Ganoderma lucidum*

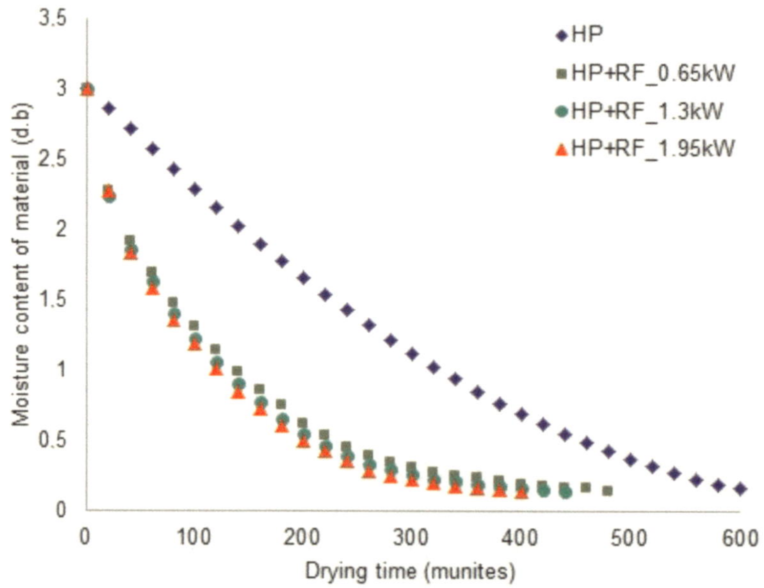


**Figure 10.**
*Drying curves of RF-assisted heat pump drying process at different RF powers.*

| No | Input drying parameter | | | Polysaccharide content (mg/g) |
|---|---|---|---|---|
| | $t_a$ (°C) | $v_a$ (m/s) | $P_{RF}$ (kW) | |
| 1 | 45 | 1.2 | 0 | 7.82 |
| 2 | 45 | 1.2 | 0.65 | 9.18 |
| 3 | 45 | 1.2 | 1.3 | 9.31 |
| 4 | 45 | 1.2 | 1.95 | 9.47 |

**Table 3.**
*Polysaccharide content of* Ganoderma lucidum *after drying.*

fluctuate faster. Thus, heat generation within *Ganoderma lucidum* becomes faster, and the moisture diffusion within *Ganoderma lucidum* occurs faster [12, 13].

### *3.4.2.2 Polysaccharide content*

The polysaccharide content of *Ganoderma lucidum* after drying is given in **Table 3**.

The data in **Table 3** shows that RF power has a significant effect on polysaccharide content of *Ganoderma lucidum* after drying. The polysaccharide content of *Ganoderma lucidum* after RF-assisted heat pump drying is considerably higher than heat pump drying. Increase in RF power retains the higher content of polysaccharide in *Ganoderma lucidum*. Generally, the reason for the degradation of polysaccharide content during drying of *Ganoderma lucidum* is due to hydrolysis, in which the polysaccharide is hydrolyzed as water is bound to the molecule [14]. RF-assisted heat pump drying process with RF heating mechanism shortens the heat treatment time, and an increase in RF power makes the linkage between water dipole molecules to be broken more easily. That can reduce the hydrolysis degree of polysaccharides.

### *3.4.2.3 Color of drying material*

Evaluation of the color change of *Ganoderma lucidum* before and after drying is conducted with X-Rite colorimeter following CIELAB scale. Fresh *Ganoderma lucidum* has the CIELAB original color value as $L_0$, $a_0$, and $b_0$. *Ganoderma lucidum* after drying has the CIELAB color value as $L^*$, $a^*$, and $b^*$. The color change index of *Ganoderma lucidum* corresponding to input drying parameters is shown in **Table 4**, in which the International Commission on Illumination (CIE) parameters as L, a, and b are measured with a colorimeter (see **Table 1**). The corresponding L value is lightness of color from 0 (black) to 100 (white); a value is degree of redness (0 to 60) or greenness (0 to −60); and b value is yellowness (0 to 60) or blueness (0 to −60). The total change in color ($\Delta E^*$) of the drying *Ganoderma lucidum* sample with reference to the original sample is calculated as Eq. (1).

The data in **Table 4** shows that the color change index as $\Delta L$, $\Delta a$, and $\Delta b$ corresponding to RF-assisted heat pump drying is considerably smaller than

| Color index | | | | | | | |
|---|---|---|---|---|---|---|---|
| **Type of sample** | **CIELAB color value** | | | **Color change index** | | | |
| | $L_0$ | $a_0$ | $b_0$ | | | | |
| Fresh samples | 47.12 | 4.11 | 18.85 | | | | |
| | $L^*$ | $a^*$ | $b^*$ | $\Delta L$ | $\Delta a$ | $\Delta b$ | $\Delta E^*$ |
| Heat pump drying ($P_{RF}$ = 0 kW) | $36.5^a$ | $6.94^a$ | $12.52^a$ | 10.62 | 2.83 | 6.33 | $12.68^a$ |
| RF-assisted heat pump drying ($P_{RF}$ = 0.65 kW) | $38.71^b$ | $5.75^b$ | $13.76^b$ | 8.41 | 1.64 | 5.09 | $9.97^b$ |
| RF-assisted heat pump drying ($P_{RF}$ = 1.3 kW) | $39.02^c$ | $5.42^c$ | $14.1^c$ | 8.1 | 1.31 | 4.75 | $9.48^c$ |
| RF-assisted heat pump drying ($P_{RF}$ = 1.95 kW) | $39.35^d$ | $5.12^d$ | $14.46^d$ | 7.77 | 1.01 | 4.39 | $8.98^d$ |

*Mean values in the same column with different letter symbols. Significant difference at significance level of 0.05.*

**Table 4.**
*The CIELAB color values of* Ganoderma lucidum.

heat pump drying and increase in RF power decreases the color change index values. Thus, the *Ganoderma lucidum* samples have retained the color better at higher RF power and at RF power of 1.95 kW, and the color of *Ganoderma lucidum* samples is nearly similar to the original brown red of fresh material samples.

### *3.4.3 Evaluation of the effect of drying air temperature*

The drying curves of RF-assisted heat pump drying process at the drying air temperature of 40, 45, and 50°C, drying air velocity of 1.2 m/s, and RF power of 1.3 kW is presented graphically in **Figure 11**.

As shown in **Figure 11**, increasing drying air temperature has a significant effect on moisture ratio; the moisture ratio is higher at higher drying air temperature. At drying air temperature of 50°C, the drying time reduces by 10% and 21% in comparison with drying air temperature of 40 and 45°C. It can be explained by the fact that the increase in drying air temperature will increase the amount of heat absorbed by material. Thus, the heating rate increases, and the moisture diffusion within *Ganoderma lucidum* occurs faster.

### *3.4.4 Evaluation of the effect of drying air velocity*

The drying curves of RF-assisted heat pump drying process at the drying air temperature of 45°C; drying air velocity of 1.2, 1.6, and 2 m/s; and RF power of 1.3 kW is presented graphically in **Figure 12**.

As shown in **Figure 12**, increasing drying air velocity makes drying time become longer. This is explained by the fact that the increase in drying air velocity will increase the drying airflow in contact with the drying material surfaces. The

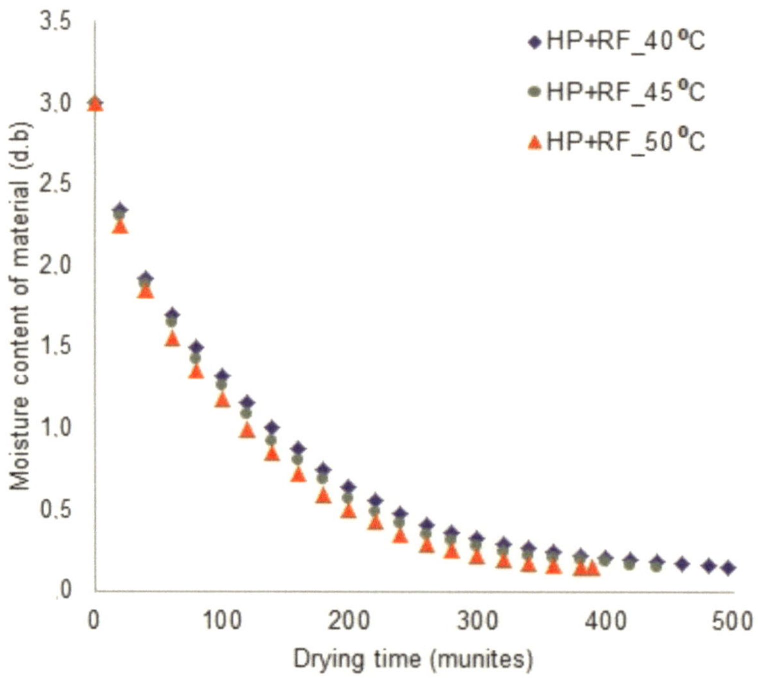


**Figure 11.**
*Drying curves of RF-assisted heat pump drying process at different drying air temperatures.*

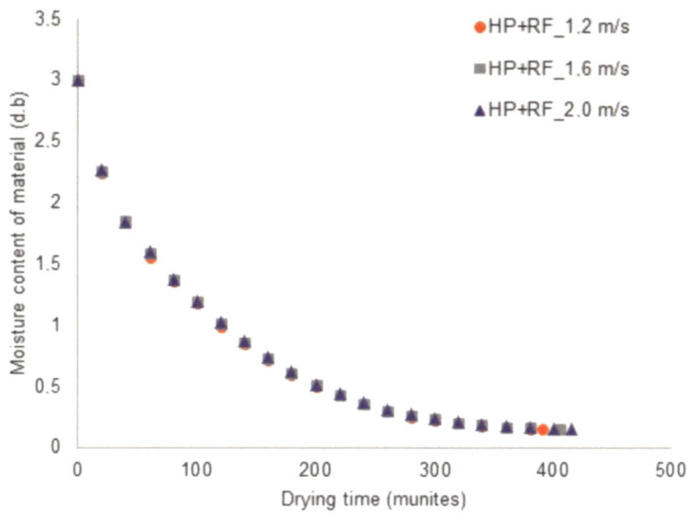


**Figure 12.**
*Drying curves of RF-assisted heat pump drying process at different drying air velocities.*

temperature of drying material is maintained at a higher level than drying air temperature during drying process by RF heating mechanism. So, when drying air comes into contact with drying material surfaces, the temperature of material surfaces will decrease that causes the average temperature of material to decrease and drying time to become longer. However, drying air velocity does not significantly affect the drying rate. The drying time corresponding to three drying air velocity values differs only about 10–15 minutes, and the drying curves shown in **Figure 12** are almost identical. The experimental results are in agreement with the previous studies of agricultural product drying [15–18].

## 4. Conclusions

Based on the calculation and design results, the RF generator has been successfully manufactured and applied in drying technology. The RF generator worked efficiently and achieved the required RF power of 3 kW and frequency of 27 MHz as designed. The drying experiment results showed that in RF-assisted heat pump drying, increase in RF power and drying air temperature increases the drying rate considerably. Meanwhile, drying air velocity does not significantly affect the drying rate. Besides, when RF power increases, the *Ganoderma lucidum* samples retain the higher content of polysaccharide and the original color better after drying.

## Nomenclature

| | |
|---|---|
| AC | alternating current |
| C | capacitor |
| $C_p$ | specific heat of drying material, J/(kg °C) |
| d.b | dry basic (kg $H_2O$/kg dry solid) |
| DC | direct current |
| f | frequency, MHz |
| $G_b$ | drying capacity, kg/batch |
| h | the height, m |

| | |
|---|---|
| HP | heat pump |
| HPLC | high-performance liquid chromatography |
| I | ampere, A |
| l | the length, m |
| L | inductor coil |
| LSD | least significant difference |
| $m_{LC}$ | mass of *Ganoderma lucidum*, kg |
| M | moisture of drying material, d.b |
| $P_{RF}$ | RF power, kW |
| Q | the heat, kW |
| r | latent heat of vaporization of moisture in material, J/kg |
| RF | radio frequency |
| t | temperature of drying material, °C |
| T | absolute temperature of drying material, °K |
| U | voltage, kV |
| w | the width, m |
| w.b | wet basic (kg $H_2O$/kg wet solid) |

**Greek symbols**

| | |
|---|---|
| $\lambda$ | thermal conductivity, W/m °C |
| $\omega$ | moisture of drying material, w.b |
| $\varepsilon$ | radiation ratio of galvanized steel |
| $\tau$ | the time, s |

**Subscripts**

| | |
|---|---|
| i | initial |
| f | final |
| W | water |
| ch | chamber |

## Author details

Nguyen Hay[1*], Le Anh Duc[1*] and Pham Van Kien[2]

1 Nong Lam University, Ho Chi Minh City, Vietnam

2 LILAMA2 International Technology College, Dong Nai, Vietnam

*Address all correspondence to:
ng.hay@hcmuaf.edu.vn and leanhduc@hcmuaf.edu.vn

## References

[1] Perumal K. Indigenous Technology on Organic Cultivation of Reishi (*Ganoderma lucidum*). Taramani, Chennai: Shri AMM Murugappa Chettiar Research Centre; 2009

[2] Ptasznik W, Zygmunt S, Kudra T. Simulation of RF assisted convective drying for seed quality broad bean. Drying Technology. 1990;**8**:977-992

[3] Jumah R. Modelling and simulation of continuous and intermittent radio frequency assisted fluidized bed drying of grains. Food and Bioproducts Processing. 2005;**83**:203-210

[4] Yanhong L et al. Heating patterns of white bread loaf in combined radio frequency and hot air treatment. Journal of Food Engineering. 2013;**116**:472-477

[5] Kaya A. Numerical analysis of a radio frequency assisted convective drying. International Journal of Scientific & Technology Research. 2015;**4**:234-241

[6] Huang Z, Zhu H, Yan R, Wang S. Simulation and prediction of radio frequency heating in dry soybeans. Biosystems Engineering. 2015;**129**:34-47

[7] Houa L, Huanga Z, Koua X, Wanga S. Computer simulation model development and validation of radio frequency heating for bulk chestnuts based on single particle approach. Food and Bioproducts Processing. 2016;**100**:372-381

[8] Jiao S, Zhu D, Deng Y, Zhao Y. Effects of hot air assisted radio frequency heating on quality and shelf-life of roasted peanuts. Food and Bioprocess Technology. 2016;**9**:308-319

[9] Zhu H, Li D, Shujun L, Wang S. A novel method to improve heating uniformity in mid-high moisture potato starch with radio frequency assisted treatment. Journal of Food Engineering. 2017;**206**:23-36

[10] Van Phu T. Calculating and Designing Drying System. Ha Noi, Vietnam: Education Publisher; 2002. p. 360

[11] Arun M. Radio frequency drying systems for the textile. In: Industry Handbook. England: Strayfield Limited; 2007. p. 352

[12] Houa L, Linga B, Wang S. Development of thermal treatment protocol for disinfesting chestnuts using radio frequency energy. Postharvest Biology and Technology. 2016;**98**:65-71

[13] Ozturk S, Kong F, Singh RK, Kuzy JD, Li C. Radio frequency heating of corn flour: Heating rate and uniform. Innovative Food Science & Emerging Technologies. 2017;**171**:234-241

[14] Chin SK, Law CL, Cheng PG. Effect of drying on crude ganoderic acids and water soluble polysaccharides content in *Ganoderma lucidum*. International Journal of Pharmacy and Pharmaceutical Sciences. 2011;**3**:38-43

[15] Celma AR, Cuadros F, Lopez-Rodriguez F. Review: Characterization of industrial tomato by products from infrared drying process. Food and Bioproducts Processing. 2009;**87**:282-291

[16] Younis M, Abdelkarim D, El-Abdein AZ. Kinetics and mathematical modeling of infrared thin—Layer drying of garlic slices. Saudi Journal of Biological Sciences. 2017;**80**:25-32

[17] Sharmaa GP, Prasadb S, Chahara VK. Moisture transport in garlic cloves undergoing microwave—Convective drying. Food and Bioproducts Processing. 2009;**87**:11-16

[18] Duc LA, Han J-W. The effects of drying conditions on the germination properties of rapeseed. Journal of Biosystems Engineering. 2009;**34**:30-36

Section 3

# Recycling and Waste Management

Chapter 5

# Auxiliary Strategies for Water Management in Industries: Minimization of Water Use and Possibility of Recycling and/or Reuse of Effluent

*Fábio Henrique de Melo Ribeiro, Yeda dos Santos Silva and Liliana Pena Naval*

## Abstract

Water management in industry by minimizing water consumption and effluent generation, reusing and/or recycling as a possibility the economy and conservation of water, energy and economic resources. The characterization of the final effluent allows evaluating how much the treatment is adequate to meet the requirements of the regulations of different countries for recycling and/or reuse and evaluated the possibility of reuse, as well as the choice of effluent treatment methods. In this case, technical, environmental and economic criteria, with a view to complying with industrial reuse regulations, should be evaluated, and a multicriteria analysis (MCA) can be adopted to classify the treatment systems applied in different reuse scenarios, made possible by the combination of multiple processes, with the use of tertiary treatment techniques. It should be noted that the potential for recycling and/or reuse of effluents generated in industry increases when effluents are separated into groups (principle of segregation of effluent streams). As a way of promoting a more sustainable model, the use of reuse systems is promising to reduce consumption, as well as reducing operating costs when treating effluents.

**Keywords:** industrial reuse, multicriteria analysis, segregation of effluents, use of effluents, water management

## 1. Introduction

Environmental pollution and the preservation of natural resources are subjects of constant presence in the world political and socioeconomic guidelines, especially the discussions related to water pollution. These debates are fueled by issues such as water scarcity [1–3], climate change pressures, disordered urban development and increased domestic waste production and industrial [4, 5].

To minimize some of these impacts, especially water scarcity, the implementation of effluent reuse programs provides direct and indirect benefits, such as the integration and sustainable use of water resources; reduction of excessive abstraction of



IntechOpen

surface and groundwater; reduction of energy consumption and environmental protection, reinforced by the restoration of rivers, marshes and lagoons [5]. With advances in effluent treatment technologies [6] and the possibility of [7, 8], water reuse presents itself as a potential source for different sectors of society, especially for the industrial sector.

The industry has also sought to improve processes in terms of sustainability, such as measures to reduce waste and effluent production, to meet international and national market requirements, and to adapt to the new scarcity scenario water resources. A number of countries practice industrial water reuse [9] the United States, Japan, and Australia have projects with a high percentage of effluent reuse from commercial and residential water and sanitation facilities [5, 10].

The establishment of targets for reuse, expressed in terms of the percentage of municipal effluents, which are treated to obtain a high quality, for an advantageous reuse have been adopted by different countries. Australia, which reused about 8% of the treated effluent in 2012, set the goal of increasing this quantity to 30% by 2030. In the case of Saudi Arabia, about 16% of the effluent was reused in of expansion to 65% by the end of 2016. Singapore, reuse 30% of the effluent and has a plan to reduce its dependence on external sources. Israel has achieved 70% reuse of domestic effluents [8].

In the case of the fish processing industry, where water is used in abundance in the various stages of production (an average of 11 $m^3$ ton of processed fish and 15 $m^3$ tons in the case of shrimp processing). The adoption of measures to reduce waste and effluent production are needed [11–14]. Under these circumstances, the use of reuse and recycling systems is promising to achieve these objectives and is important for achieving sustainable management [8]. In this industry, the large amount used leads to an increase in the volume of effluents generated, which, if not treated properly, lead to different impacts [7, 11–13, 15].

Evaluating alternatives for treatment of effluents capable of meeting technical, environmental and economic criteria implies the feasibility of reducing effluents and improving quality. The adoption of technologies and procedures that reduce the amount of water used. As well as the increase in reuse can characterize the implantation of cleaner production technologies in the industries, which not only confers the reduction of the direct and indirect costs of the process through the management of the consumption of water, energy and raw material used, as well as the efficiency of the enterprise.

## 2. Quality and requirements established for the practice of industrial reuse

As for the water consumption in the fish processing industries, it is known that the greater use is concentrated in the washing and cleaning steps. However, volumes used for the storage and refrigeration of fishery products should be considered for reuse [16], both before and during processing, as an important lubricant in the various stages of fish handling [16, 17]. As well as in waste management, which consists of scales, meat, bones, cartilage and viscera [11] and of the effluents characterized by high organic load and salts, which result in higher volumes of total suspended solids, biochemical oxygen demand (BOD) and chemical oxygen demand (COD) [11, 16, 18, 19] reducing the quality of these effluents. They are still rich in nitrogen and phosphorus, which when discharged can lead to eutrophication [12].

Other factors to be considered in production that will influence the effluent characteristics are the types of fish to be processed, the water supply system used

| Concentration limits | Application | Country |
|---|---|---|
| Fecal coliform <200 NPM/100 mL<br>pH between 6 and 8<br>Turbidity <5 NTU<br>Free residual chlorine between 0.5 and 1.5 mg/L | Class 1 water: used for car washing, direct contact of users with water, aerosol aspiration | Brazil |
| Fecal coliform <500 NPM/100 mL<br>Turbidity <5 NTU<br>Free residual chlorine >0.5 mg/L | Class 2 water: used for washing floors and pavements, watering gardens, maintaining lakes and canals for landscape purposes, except fountains | |
| Fecal coliform <500 NPM/100 mL | Class 3 water: used for flushing toilets | |
| Fecal coliform <500 NPM/100 mL<br>Turbidity <10 NTU<br>Dissolved oxygen >2 mg/L | Class 4 water: reuse in orchards, cereal, fodder, cattle pastures and other crops through surface drainage or specific irrigation systems. | |
| TSS: 35 mg/L<br>Turbidity: 15 NTU<br>*Escherichia coli*: $10^4$ UFC/100 mL<br>*Legionella* spp.:100 UFC/L | Cleaning process, but not for the food industry | Spain |
| TSS: 35 mg/L<br>*Escherichia coli*: $10^3$ UFC/100 mL<br>Nematode eggs: 1 eggs/10 mL<br>*Legionella* spp.: 100 UFC/L | Processing and washing water in the food industry | |
| TSS: 5 mg/L<br>Turbidity: 1 NTU<br>*Escherichia coli*: 0 UFC/100 mL<br>Nematode eggs: 1 eggs/10 mL | Cooling towers and evaporative condensers | |
| BOD ≤30 mg/L<br>Thermotolerant coliforms ≤200/100 mL<br>pH between 6 and 9<br>TSS ≤30 mg/L<br>Minimum residual chlorine 1 mg/L | Cooling without recirculation | USA |
| BOD ≤30 mg/L<br>Thermotolerant coliforms ≤200/100 mL<br>pH between 6 and 9<br>TSS ≤30 mg/L<br>Minimum residual chlorine 1 mg/L | Cooling towers (variables depend on the rate of recirculation) | |
| *Escherichia coli* ≤200 UFC/100 mL (average) | Cooling water | Greece |
| pH between 6 and 8.5<br>BOD: ≤10 mg/L (in 80% of samples)<br>TSS: ≤10 mg/L (in 80% of samples)<br>Turbidity ≤2 NTU<br>*Escherichia coli*: ≤5 UFC/100 mL (80% of samples), ≤50 (in 95% of samples) | Use of single reticulated cooling water, cooling water for boilers, process water. | |

*TSS, total suspended solids; BOD, biochemical oxygen demand; NTU, turbidity unit; UFC, colony formation unit. Brazilian NBR Technical Standards 13969: 1997 [24], European Standards: Spain, Royal Decree 1620, [25] and Greece, Ministerial Decree [26], American Guidelines [8].*

**Table 1.**
*Reuse and recycling quality requirements established by standards and regulations for industrial reuse.*

and the volume of effluent generated [16, 18, 20]. The occurrence of these variables in the operational conditions makes it difficult to plan a single treatment unit capable of meeting the necessary requirements for all types of effluents produced in this type of industry.

When it is intended to employ water reuse systems in meat product industries, account should be taken of the limitation imposed by the regulations. Reuse in these industries is generally restricted to direct or indirect reuse for operations where water does not come into contact with the product being processed or, in some situations, with whom it is handled. There are also other barriers to the large-scale operationalization of these systems, such as insufficient policies to support the reuse of reclaimed water; lack of public awareness and acceptance; failures in risk management systems, among others [6, 8]. However, each industrial plant is unique, with size and quality of different effluents, therefore, generalizations about the use and effluent characteristics are difficult to measure, making treatment complex.

Another barrier is the environmental regulations, which focus on the discharge of effluents into the water bodies, not being considered, in most of these documents, the necessary criteria for reuse and recycling [21]. However, there are efforts by several countries. In Europe, the countries with more specific reuse regulations are Greece, Spain and Portugal, and have applied in different reuse modalities. Italian regulations also describe urban, agricultural and industrial uses, but industrial use is permitted if there is no direct contact with food [5]. In the United States, regulations are developed according to the criteria of each state. In the case of Australia, government agencies initiated a reform in water management in 1994, when measures were adopted to use alternative water sources and the development of guidelines for obtaining recycled drinking water [22, 23]. In number of reuse projects, Japan leads, with approximately 1800; followed by Australia with more than 450; then Europe, with about 200; the Middle East, with more than 100; Latin America, over 50, and Sub-Saharan Africa, with just over 20 [5].

Despite the legal limitations of reuse, countries have been regulating the practice (**Table 1**). In these places, reuse water is mainly applied to urban uses and agricultural irrigation. Although effluent reuse is a widespread and widely applied practice, it is necessary to remember that the accomplishment of treatment to suit the requirements of the next use or to the related regulations is indispensable.

## 3. Analysis of potential effluent reuse and recycling

The industrial sector has adopted water reuse programs (**Table 2**), as a tool for the economy, for sustainability, and for the preservation of water resources. In order to comply with the regulations for industrial reuse and potability, joint systems of treatments are required. However, conventional effluent treatment is not suitable for the application of the effluent treatment, since the use of a less expensive technology for the treatment of effluents when the reuse option adopted is less restrictive [8, 24]. As well as the reuse and/or recycle systems when it comes to the food industry, since it is necessary to meet the specific criteria [16, 27, 28]. Advanced treatment techniques capable of removing high levels of pollutants should be used [29].

The choice of treatment technologies that best fits the reality of each industry does not depend exclusively on the level of removal to be achieved, since other technical, economic and environmental criteria also influence decision-making. In order to establish which treatment levels are adequate, tools capable of evaluating the technologies for applications in reuse projects can be employed. Compensatory models are an example; since they allow achieving results closer to what would be the ideal result, because they are more demanding in assessing the advantages and disadvantages of each attribute, which characterizes a multicriteria analysis (MCA). These models can be divided into three subgroups: (i) scoring models (such as simple additive weighting); (ii) compromise models (such as TOPSIS); and (iii) concordance models (such as ELECTRE and PROMETHEE) [35].

| Product or stage of processing | Type of water | Subsequent use | Reference |
|---|---|---|---|
| Beef processing | | | |
| Beef processing | Shower water from chiller | Reuse as warm water for cleaning and as water of constitution of boiler | [30] |
| Beef processing | | | |
| | Final effluent | Reuse in operation that does not require low concentration<br>Total solid suspended (SST) and turbidity | [31] |
| Poultry processing | | | |
| Poultry processing | Pre-chiller water, effluent from gutter gutter; cooling chamber water and thawing; filter washing water | Recycle | [32] |
| Fish processing | | | |
| Processing of crustaceans | Cooling water from crustaceans after cooking | Direct reuse in the crustacean cleaning step before cooking | [33] |
| Fish processing | Process water from: vacuum or centrifugal pumping, evaporation (film evaporator or conventional) and drying (direct flame or steam) | Recycling, protein separation and subsequent incorporation into the processing | [34] |

*TSS, total suspended solids.*

**Table 2.**
*Industrial recycling and reuse of treated effluents in food processors.*

The PROMETHEE is a non-parametric method of classification, which uses the principle of superior classification to formulate a ranking of alternatives, suitable for problems in which a finite number of alternatives must be classified in relation to several, sometimes-contradictory criteria [36]. The PROMETHEE approach has the advantage of being easier to use and less complex than the ELECTRE approach, although they are part of the same principles of agreement. For this reason, its application in solving environmental problems is increasing [3, 35].

For the implementation of PROMETHEE, it is necessary to define the weights of the criteria adopted and the preference functions of the decision maker when comparing the contribution of the alternatives in terms of each separate criterion. ELECTRE is a method of overcoming based on the agreement analysis. The main advantage is that it takes account of uncertainties and inaccuracies.

By optimizing the way in which the industries treat the effluents, a reduction of the operational costs of the plant can be obtained, besides minimizing the generation and the volume of effluents, without sacrificing the value or quality of the product [37]. The multicriteria analysis can subsidize the choice of the technology that satisfies the most possible criteria (objective and subjective), considering aspects competing in the decision of the managers of these types of establishments. Among the alternatives of effluent treatment systems for the fish processing industry, it is possible to choose the technologies that have the highest levels of removal (**Table 3**), capable of producing reuse waters with higher quality, or by better meeting the criteria of greater relevance [38].

From the removal rate obtained by the different treatment systems, multicriteria analysis (AMC) can be employed to support the decision on the choice of

| Treatment | Parameters | Removal | Reference |
|---|---|---|---|
| Coagulation/Flocculation with $FeCl_3$ | SST<br>BOD<br>COD<br>Oils and greases | 95.4%<br>89.3%<br>87.5%<br>92% | [39] |
| Rotary bioreactor | COD | 98% | [40] |
| Discontinuous mixed reactor and compact filter reactor | Ntotal<br>Dissolved organic carbon | 99.9%<br>88% | [41] |
| Bioreactor and ultrafiltration by membranes | COD | 92% | [42] |
| Bioreactor; coagulation/flocculation/sedimentation; microfiltration by membranes | COD<br>Dissolved solids<br>Ntotal<br>Ptotal | 100%<br>100%<br>93%<br>100% | [15] |
| Sedimentation/flotation; coagulation/flocculation; biological treatment by activated sludge process; sand filter filtration; reverse osmosis and UV disinfection | Dissolved organic carbon<br>Oils and greases<br>SST<br>Anions and cations heterotrophic bacteria | 99.9%<br>99.8%<br>98.4%<br>96%<br>100% | [13] |

*TSS, total suspended solids; BOD, biochemical oxygen demand; COD, chemical oxygen demand; Ntotal, total nitrogen; Ptotal, total phosphorus; UV, ultra violet. Source: [43].*

**Table 3.**
*Advanced treatment techniques for removal of high levels of pollutants.*

wastewater treatment systems for reuse, considering the economic, technological and environmental criteria.

The AMC tool was employed to determine the best wastewater treatment technology from fish processing industries. The Visual PROMETHEE 1.4 program (implementation software of both the PROMETHEE method and the GAIA method) was used.

Economic [construction cost (CC) and operation and maintenance cost (CO&M)], technological [pollutant removal capacity (CRP), system complexity (COMP) and specialized MO (MOE)] and environmental (potability) aspects were adopted. [(PO), energy consumption (EC) and odors (OD)] [38].

It was postulated that the best system comprises efficient and low cost treatment and it was admitted that the economic and environmental criteria have importance and greater weight in the analysis. When considering obtaining a wastewater for potable reuse, deployment costs and removal efficiency were prioritized. While operation and maintenance costs were of intermediate importance, followed by the other criteria in order of importance. If potable reuse was not necessary, its weight was redistributed, prioritizing cost and removal efficiency criteria.

To analyze effluent compliance for reuse, the following standards and regulations were adopted (**Table 1**): Brazilian NBR Technical Standards 13969: 1997 [24], European Standards: Spain, Royal Decree 1620, [25] and Greece, Ministerial Decree [26], American Guidelines [8].

When the intended reuse was drinking, the Brazilian Ministry of Health (MS Ordinance No. 2914 [64]) was used, as well as to evaluate the potential for reuse of effluents in the most restrictive activities, such as preparation, handling and disposal fish packaging in processing industries.

For the use of potable reuse, from the effluent generated in the facilities of fish processing industries, the potability criterion (PO) was considered to be of greater relative importance, with a higher weight valuation than the others, given the restrictions imposed by the use itself and by rules and regulations.

Among the alternatives of effluent treatment systems for the fish processing industry, the technologies that presented the highest levels of pollutant removal were chosen (**Table 3**).

After processing the data by the AMC program used, it can be verified that the alternative that best meets the criteria listed for the importance weighting adopted was the effluent treatment system composed by the following technologies: bioreactor; coagulation/flocculation/sedimentation; microfiltration by membranes (Bio + Coag/floc/sed + Memb) [15], followed by sedimentation/flotation systems; coagulation/flocculation; biological treatment by activated sludge process; sand filter filtration; reverse osmosis and UV disinfection (Sed/Flot + Coag/floc + Sludge + Filtr + OsmRev + UV) [13].

It is noteworthy that the alternative Bio + Coag/floc/sed + Memb presented better overall performance, by better meeting the most relevant criteria adopted for drinking reuse. The criteria that most influenced the decision axes (**Figure 1**) were pollutant removal efficiency (CRP) and potability (PO).

For the reuse of non-potable water in less restrictive activities associated with the fish processing industry, such as use in water sanitation facilities, floor washing, garden irrigation and cooling and heating systems, potability requirements were not considered. Therefore, the valuation of the weights presented a redistribution of importance, prioritizing the criteria construction cost (CC), operation and maintenance cost (CO&M) and pollutant removal capacity (CRP).

As alternatives for effluent treatment systems, the same technologies were adopted when considering potability. For this case, the alternative that best meets the listed criteria for the importance weighting adopted was the effluent treatment system proposed by Fahim et al. [39]: coagulation/flocculation with $FeCl_3$ (Coag/Floc), followed by the systems proposed by Artiga et al. [42]: bioreactor and ultrafiltration by membranes (Bio + Memb).

The reason for the alternative proposed by Fahim et al. [39] presented the best overall performance, by meeting the most relevant criteria adopted: construction cost (CC) and operation and maintenance cost (CO&M) and pollutant removal capacity (CRP). For and Artiga et al. [42] were the COMP, CC and EC criteria (**Figure 2**).

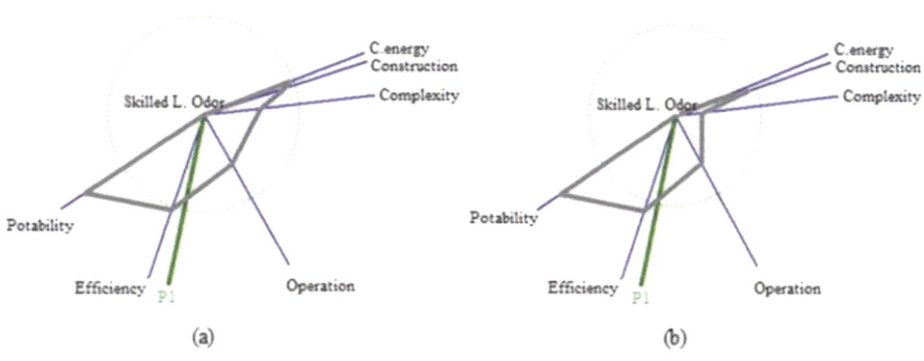


**Figure 1.**
*Behavior of treatment alternatives proposed by Queiroz et al. [15], (a) bioreactor; coagulation/flocculation/sedimentation; microfiltration by membranes (Bio + Coag/floc/sed + Memb), followed by the systems proposed by Cristóvão et al. [13]: sedimentation/flotation; coagulation/flocculation; biological treatment by activated sludge process; sand filter filtration; reverse osmosis and UV disinfection (Sed/Flot + Coag/floc + Sludge + Filter + OsmRev + UV).*

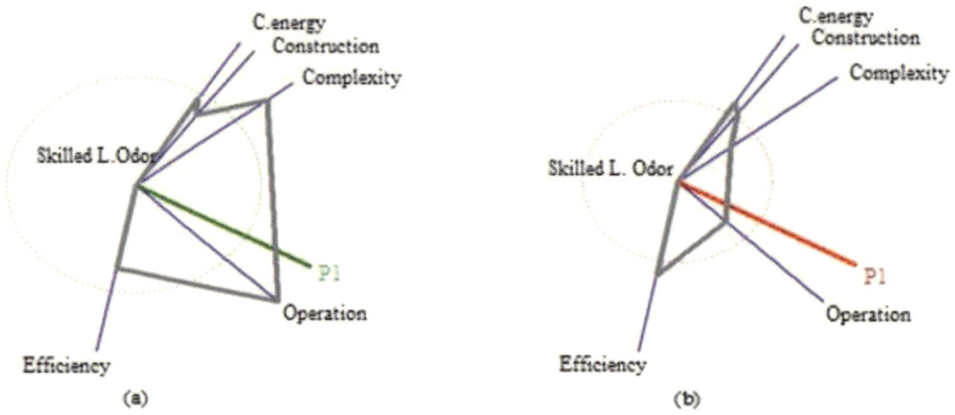


**Figure 2.**
*Behavior of Fahim et al. [39] (a) coagulation/flocculation with $FeCl_3$ (Coag/Floc), Artiga et al. [42] (b), Queiroz et al. [15] bioreactor and ultrafiltration by membranes (Bio + Memb) and their decision axes.*

| Technician employed | % Theoretical adopted | References |
|---|---|---|
| **Total solids** | | |
| Screen | 31% a 60% | [46, 53] |
| Linked screen with fil | 40% a 70% | [50] |
| and catchment area | 100% | [54] |
| Sieving conjugated with microfiltration, ultrafiltration, nanofiltration and reverse osmosis | 80% a 90% | [49] |
| **Organic matter** | | |
| Screens | 25% a 60% | [29] |
| Rotary filter | 15% | [45] |
| Rotary sieve | 25% | [45] |
| Nanofiltration conjugated prefiltration | 56%. | [54] |
| Ultrafiltration | 36% | [54] |
| Nanofiltration | 60% a 80% | [54] |
| Dissolved air flotation | 30% a 90% | [29, 55] |
| Coagulation-flotation | 90% | [56] |
| Reverse osmosis | 97,50% | [54] |
| **Oils and greases** | | |
| Membrane filtration associated with electrocoagulation | 65% | [57] |
| Ceramic membrane and electrocoagulation | 50% | [57] |
| Ceramic membrane | 2% | [57] |
| Dynamic membrane | 10% | [57] |
| Flotation | 37% e 63% | [49] |
| Screen | 10% a 20% | [49] |

*Source: [58].*

**Table 4.**
*Segregation techniques and removal percentages achieved.*

Even if the desired level of pollutant removal is reached, the use of clean technologies, which promote green innovation, together with the production process, should favor the sustainability of product transformation [44, 45]. Complex or simple technological investments, such as segregation of effluent, in processing contribute to cleaner production [44, 46, 47]. Segregation facilitates the treatment of the generated effluents [17] and can occur in the processing, through optimizations added to the production line.

In the case of the fish processing industry, several alternatives can be adopted such as alteration in the cutting machine; adjustment in the mechanized filleting machine; inclusion of waste separation ramp; [48] (**Table 4**). Allied operations are considered the minimization of waste generation, such as sieving; filtration; [49, 50] which reduce between 30% and 80% of the solid residues originated during fish processing [51, 52].

Studies of the valuation of by-products of fish processing indicate that these can be used in the elaboration of new products, with low raw material and production costs, increasing the industry profit and reducing the environmental impact caused [17]. Among the alternatives for the reuse of waste generated by fish processing are the use for animal and human consumption and for biodiesel generation, which may contribute to the establishment of a sector committed to environmental issues [59–63].

## 4. Conclusions

The precise characterization of the effluents, including the daily volumes, flow rates and associated pollutant load, is fundamental for an efficient design of the treatment systems. The determination of the performance requirements of the treatment systems depends directly on a detailed assessment of the quality of the effluents to be treated.

The choice of the treatment system to be used with a view to reuse, capable of guaranteeing the project's profitability and sustainability, is not a simple decision process, depending on the number of possible alternatives and criteria to be evaluated (such as economic, technical, environmental and social). In order to choose the most appropriate technologies for the treatment of effluents, it is necessary to define the intended destination, either for discharge into the water sources or for the application in reuse and/or recycling systems. Based on related regulations, the available technologies can be related to the levels of removal required.

## Acknowledgements

The authors thank the National Council for Scientific and Technological Development for the productivity grant (Process 40.3291/2016-0).

## Conflict of interest

The authors declare no conflict of interest.

## Author details

Fábio Henrique de Melo Ribeiro, Yeda dos Santos Silva and Liliana Pena Naval*
Federal University of Tocantins (UFT), Palmas, Tocantins, Brazil

*Address all correspondence to: liliana@uft.edu.br

## References

[1] EPA. Guidelines for Water Reuse. Washington, DC: Environmental Protection Agency. [Internet]. 2004. Available from: https://www3.epa.gov/region1/npdes/merrimackstation/pdfs/ar/AR-1530.pdf [Accessed: 21 April 2017]

[2] Arevalo J, Ruiz LM, Parada-Albarracín JA, González-Pérez DM, Pérez J, Moreno B, et al. Wastewater reuse after treatment by MBR. Microfiltration or ultrafiltration. Desalination. 2012;**299**:22-27

[3] Sadr S et al. Appraisal of membrane processes for technology selection in centralized wastewater reuse scenarios. Sustainable Environment Research. 2013;**23**:69-78

[4] Almeida CMVB et al. Identifying improvements in water management of bus-washing stations in Brazil. Resources, Conservation and Recycling. 2010;**54**:821-831

[5] Alcalde-Sanz L, Gawlik BM. Water Reuse in Europe Relevant Guidelines, Needs for and Barriers to Innovation. Luxemburg, 48 p: Publications Office of the European Union; 2014

[6] Yi L, Jiao W, Chen X, Chen W. An overview of reclaimed water reuse in China. Journal of Environmental Sciences. 2011;**23v**:1585-1593

[7] Hidalgo D, Irusta R, Martinez L, Fatta D, Papadoupolos A. Development of a multi-function software decision support tool for the promotion of the safe reuse of treated urban wastewater. Desalination. 2007;**215**:90-103

[8] EPA. Environmental Protection Agency (USA), 2012. Guidelines for Water Reuse. EPA/600/R-12/618. Washington, D.C., 643 p. Available from: http://nepis.epa.gov/Adobe/PDF/P100FS7K.pdf [Accessed: 12 April 2019]

[9] Angelakis A, Gikas P. Water reuse: Overview of current practices and trends in the world with emphasis on EU states. Water Utility Journal. 2014;**8**:67-78

[10] Leung RWK, Li DCH, Yu WK, Chui HK, Lee TO, Van Loosdrecht MC, et al. Integration of seawater and grey water reuse to maximize alternative water resource for coastal areas: The case of the Hong Kong International Airport. Water Science and Technology. 2012;**65**:410-417

[11] Anh P, Dieu T, Mol A, Kroeze C, Bush S. Towards eco-agro industrial clusters in aquatic production: The case of shrimp processing industry in Vietnam. Journal of Cleaner Production. 2018;**19**:2107-2118

[12] Muthukumaran S, Baskaran K. Organic and nutrient reduction in a fish processing facility—A case study. International Biodeterioration and Biodegradation. 2013;**85**:563-570

[13] Cristóvão RO, Botelho C, Martins R, Loureiro J, Boaventura R. Fish canning industry wastewater treatment for water reuse—A case study. Journal of Cleaner Production. 2015; [S.I.], n. 87, p. 603-612

[14] Souza MA. Eficiência do processo de ultrafiltração seguido de biodigestão anaeróbia no tratamento de efluente de frigorífico de tilápia [thesis]. São Paulo: Universidade Estadual Paulista, São Paulo; 2010

[15] Queiroz MI, Hornes M, Manetti A, Zepka L, Jacob-Lopes L. Fish processing wastewater as a platform of the microalgal biorefineries. Biosystems Engineering. 2013;**115**:195-122

[16] Chowdhury P, Viraraghavan T, Srinivasan A. Biological treatment processes for fish processing

wastewater—A review. Bio/Technology. 2010. [S.I.], n. 101, p. 439-449

[17] Arvanitoyannis ISE, Kassaveti A. Fish industry waste: Treatments, environmental impacts, current and potential uses. International Journal of Food Science and Technology. 2008;**43**:726-745

[18] Cristóvão RO et al. Chemical and biological treatment of a fish canning wastewater. International Journal of Bioscience, Biochemistry and Bioinformatics. 2012;**2**(4):237-242

[19] Cristóvão RO et al. Chemical oxidation of fish canning wastewater by fenton's reagent. Journal of Environmental Chemical Engineering. 2014; [S.I.], n. 2, p. 2372-2376

[20] Alexandre VMF et al. Performance of anaerobic bioreactor treating fish-processing plant wastewater pre-hydrolyzed with a solid enzyme pool. Renewable Energy. 2011; [S.I.], n. 36, p. 3439-3444

[21] Oliveira-Esquerre KP et al. Taking advantage of storm and waste water retention basins as part of water use minimization in industrial site. Resources, Conservation and Recycling. 2011; [S.I.], n. 55, p. 316-324

[22] Radcliff EJ. Water recycling in Australia—During and after the drought. Environmental Science: Water Research & Technology. 2015; [S.I.], 1v, n. 5, p. 554-562

[23] New South Wales Government (NSW). NSW Water reuse guideline for food businesses. NSW Food Authority, Sydney. [Internet]. 2008. Available from: https://www.environment.nsw.gov.au/topics/water/water-quality/protecting-and-managing-water-quality/waterway-health [Accessed: 11 May 2017]

[24] ABNT (Associação Brasileira de Normas Técnicas). Tanque sépticos—Unidades de tratamento complementar e disposição final dos efluentes—Projeto, construção e operação—NBR 13969. Rio de Janeiro; 1997

[25] ESPAÑA—Real Decreto 1620/2007 de 7 Diciembre, por el que se establece el regímen jurídico de la reutilización de las aguas depuradas. BOE n. 294. pp. 50639-50661. Madrid [Internet]. 2007. Available from: https://www.boe.es/eli/es/rd/2007/12/07/1620 [Accessed: 12 May 2017]

[26] Joint Ministerial Decree (JMD) 145116/2011: Definition of measures, conditions and procedure for wastewater reuse. Greek Government Gazette 354B. [Internet]. 2011. Available from: https://rm.coe.int/16805c7703 [Accessed: 21 April 2017]

[27] EPA. Guidelines for Water Reuse. Washington, D.C. Environmental Protection Agency. [Internet]. 2012. Available from: https://www3.epa.gov/region1/npdes/merrimackstation/pdfs/ar/AR-1530.pdf [Accessed: 21 April 2017]

[28] Norton T, Misiewicz P. Ozone for water treatment and its potential for process water reuse in the food industry. In: O'Donnell, C. et al. (Ed.). Ozone in Food Processing. 1. ed. Oxford: Blackwell Publishing Ltd., 2012.p. 177-200, cap. 11

[29] Mittal GS. Treatment of wastewater from abattoirs before land application—A review. Bioresource Technology. 2006;**97**(9):1119-1135

[30] Mavrov V, Chmiel H, Bélières E. Spent process water desalination and organic removal by membranes for water reuse in the food industry. Desalination. 2001;**138**:65-74

[31] Wu J, Doan H. Disinfection of recycled red-meat-processing wastewater by ozone. Journal

of Chemical Technology and Biotechnology. 2015;**80**:828-833

[32] Matsumura EM, Mierzwa JC. Water conservation and reuse in poultry processing plant—A case study. Resources, Conservation and Recycling. 2008;**52**:835-842

[33] River L, Aspe E, Roeckel M, Mart MC. Evaluation of clean technology processes in the marine products processing industry. Journal of Chemical Technology and Biotechnology. 1998;**73**:217-226

[34] Roeckel M, Aspe E. Achieving clean technology in the fishmeal industry by addition of new process step. Journal Chemical Biotechnology. 1996;**67**:96-104

[35] Kalbar P, Karmakar S, Asolekar S. Selection of an appropriate wastewater treatment technology: A scenario-based multiple-attribute decision-making approach. Journal of Environmental Management 2012; [S.I.], n. 113, p. 158-169

[36] Brans JP, Vincke PH, Mareschal B. How to select and how to rank project the PROMETHEE method. European Journal of Operational Research. 1986;**24**:228-238

[37] EPA. Emerging Technologies for Wastewater Treatment and In-Plant Wet Weather Management. Washington, D.C.: Office of Wastewater Management U.S. Environmental Protection Agency; 2013

[38] Ribeiro MF, Naval LP. Reuse alternatives for effluents from the fish processing industry through multi-criteria analysis. Journal of Cleaner Production. 2019;**19**:59-65

[39] Fahim FA et al. Evaluation of some methods for fish canning wastewater treatment. Water, Air, & Soil Pollution. 2001;**127**:205-226

[40] Najafpour GD, Zinatizadeh AAL, Lee LK. Performance of a three-stage aerobic RBC reactor in food canning wastewater treatment. Biochemical Engineering Journal. 2006, 2006;**30**:297-302

[41] Huiliñir C et al. Simultaneous nitrate and organic matter removal from salmon industry wastewater: The effect of C/N ratio, nitrate concentration and organic load rate on batch and continuous process. Journal of Environmental Management 2001; [S.I.], n. 101, p. 82-91

[42] Artiga P et al. Use of a hybrid membrane bioreactor for the treatment of saline wastewater from a fish-canning factory. Desalination 2008; [S.I.], n. 221, p. 518-525

[43] Ribeiro FHM, Naval LP. Technologies for wastewater treatment from the fish processing industry: Reuse alternatives. The Brazilian Journal of Environmental Sciences. 2017;**46**:130-144

[44] Bar ES. A case study of obstacles and enablers for green innovation within the fish processing equipment industry. Journal of Cleaner Production. 2015;**90**(1):234-243

[45] COWI, Consulting Engineers and Planners AS, Denmark. Cleaner production assessment a meat processing. Paris. UNEP—United Nations Environment Programme, Division of Technology, Industry and Economics. (ver. 2008) [Internet]. Copenhagen: Danish Environmental Protection Agency; 2008;**1**:83. Available from: https://www3. https://digitallibrary.un.org/record/441711.pdf [Accessed: 25 April 2017]

[46] Bezama A. Evaluation of the environmental impacts of a cleaner production agreement by frozen fish facilities in the Biobío Region,

Chile. Journal of Cleaner Production. 2012;**26**:95-100

[47] Denham FC et al. Environmental supply chain management in the seafood industry: Past, present and future approaches. Journal of Cleaner Production. 2015:82-90

[48] Watson R. Trials to Reduce Water and Effluent Charges in Fish Processing. The Sea Fish Industry Authority. Seafish Report N.SR; 2013;**1**:541

[49] Colic M et al. Case study: Fish processing plant wastewater treatment. Clean Water Technology, Inc. 2007:1-27. DOI: 10.2175/193864707787781557

[50] FAO. The State of World Fisheries and Aquaculture: opportunities and challenges. Rome: FAO Fisheries and Aquaculture Department. p. 233. [Internet]. 2004. Available from: http://www.fao.org/3/i9540en/i9540en.pdf [Accessed: 01 June 2018]

[51] Johanson K. Review of new segregation tester method by Dr. Kerry Johanson, P.E. Powder Technology. 2014;**257**:1-10

[52] Sutherland K. Filter media guidelines: Selecting the right filter media. Filtration & Separation. 2011;**48**(3):21-22. DOI: 10.1016/S0015-1882(11)70195-1

[53] Almandoz MC et al. Composite ceramic membranes from natural alumina silicates for microfiltration applications. Ceramics International. 2015;**4**(41):5621-5633

[54] Gebreyohannes AY, Mazzei R, Gior L. Trends and current practices of olive mill wastewater treatment: Application of integrated membrane process and its future perspective. Separation and Purification Technology. 2016;**162**

[55] Bustillo-Lecompte CF, Mehrvar M. Slaughterhouse wastewater characteristics, treatment, and management in the meat processing industry: A review on trends and advances. Journal of Environmental Management. 2015;**161**(15):287-302, set. 2015

[56] Lefebvre O, Moletta R. Treatment of organic pollution in industrial saline wastewater: A literature review. Water Research. 2016;**40**(20):3671-3682

[57] Yang T et al. Improving performance of dynamic membrane assisted by electrocoagulation for treatment of oily wastewater: Effect of electrolytic conditions. Desalination. 2016;**363**:134-143

[58] Silva YS, Naval LP. Segregation of solid waste from a fish-processing industry: a sustainable action. Rev. Ambiente e agua. 2018;**13**(2):e2155. DOI: 10.4136/ambi-agua.2155

[59] Alonso AA et al. Contributing to fisheries sustainability by making the best possible use of their resources: The be fair initiative. Trends in Food Science and Technology. 2010;**21**(0):569-578

[60] Adeoti IA and Hawboldt K. Hawboldt, 2014. A review of lipid extraction from fish processing by-product for use as a biofuel. Biomass and Bioenergy 63, abr. 2014, p. 330-340

[61] Costa JF. Biodiesel production using oil from fish canning industry wastes. Energy Conversion and Management. 2013;**74**:17-23

[62] Feltes MCM et al. Alternativas para a agregação de valor aos resíduos da industrialização de peixe. Revista Brasileira de Engenharia Agrícola e Ambiental. 2010;**14**(6):669-677

[63] Jayasinghe P, Hawboldt K. A review of bio-oils from waste biomass: Focus on fish processing waste. Renewable

and Sustainable Energy Reviews. 2012;**1**:798-821

[64] Brasil, 2011. Brasil. Ministério da Saúde. Portaria n. 2.914, dispõe Sobre Procedimentos Para o Controle e Monitoramento da Qualidade da água Para Consumo Humano e padrão de água potável. Jornal Oficial da União, Brasília; 2011

Chapter 6

# The Thermal Waste Treatment Plant in Kraków, Poland: A Case Study

*Dariusz Sala and Bogusław Bieda*


## Abstract

The thermal waste treatment plant (TWTP) in Kraków (eco-incinerator) was created as a response to the energy and ecological needs of Kraków as part of the project "Municipal Waste Management Program in Krakow." The TWTP is able to process 220,000 tons of municipal waste during the year. Estimated values of the 65,000 MWh of electricity and 280,000 MWh of heat are produced as a result of the waste combustion. The energy obtained by way of the thermal transformation process is largely organic and renewable. The TWTP is equipped with a state-of-the-art exhaust purification system that meets strict emission standards for air. The emission standards will meet the requirements the Ordinance of the Minister of the Environment of November 4, 2014 on emission standards for certain types of installations, sources of fuel combustion and devices for incineration or co-incineration of waste (Journal of Laws of 2014, item 1546, including further amendments). The cleaning process takes place in the exhaust aftertreatment process and is based on the following steps: (*i*) denitrification of exhaust gases, (*ii*) flue gas cleaning by means of a semi-dry method and (*iii*) dust extraction. As the project's general contractor was POSCO E&C from South Korea.


**Keywords:** Poland, emission standards, waste incineration, exhaust aftertreatment process, heat, electricity

## 1. Introduction

The term "thermal treatment" is used to describe a range of technologies that use heat to degrade the constitution of solid matter. These include incineration and its variations, as well as advanced thermal conversion (ATC) technologies such as pyrolysis and gasification [1, 2]. Incineration of waste has been practiced for more than a century in the industrialized world [3].

For many people, thermal treatment technologies for waste management represent an image of hell on earth [2]. In line with Moberg et al. [4], waste is generated as a consequence of most of our daily activities. Waste incineration is often (but not always) the preferable choice when incineration substitutes landfill disposal of waste [5]. According to Hauck et al. [6], modern WTE facilities process approximately 13% of the total municipal waste in the United States. There is potentially more than 16,000 MW of electric power that currently a "missed opportunity" in the United States alone. The average gross and net electrical power generation of

WTE facilities has increased over the past decade to approximately 550 kWh per net ton of waste processed, assuming a typical municipal solid waste (MSW) heating value of 5000 Btu per pound (deg. F) (*Note: 5000 Btu per pound (deg. K = 20,934 kJ/kg (deg. K) = 20.934 MJ/kg = 20.934 GJ/ton*) [6].

The first incinerators were developed in the United kingdom in the last part of the nineteenth century [3]. Germany introduced the technology in Hamburg in 1895 followed by Brussels, Stockholm and Zurich in 1904. British technology was used for the first plants in other parts of Europe. This includes Denmark where the first incinerators were constructed in the Copenhagen area in 1903 by the British company Hughes & Stirling and steam boilers from Babcock and Wilcox [3]. More reading on the history of waste incineration is available in Chandler et al. [7] and Kleis and Dalager [8].

The list of waste conversion technologies includes advanced combustion, anaerobic digestion, catalytic depolymerization, fermentation, gasification and pyrolysis [6]. Several cited technologies are currently being evaluated in various stages of testing, with funding provided by the US Department of Energy and private investors, and by 2020, it is possible that several of the presented above waste conversion technologies will advance to commercial status (HAUCK). The WTE industry is still evolving and be used and implemented into a municipal waste management process.

As reported by Chromec and Ferraro [9] in December 2007, the United Nations Framework Convention on Climate Change (UNFCCC) took place in Bali. Key mitigation technologies in the waste sector, landfill gas (LFG) methane recovery, waste incineration with energy recovery, composting of organic waste, controlled waste water treatment, recycling and waste minimization, biocovers and biofilters to optimize methane oxidation have been proposed. In the presented above mitigation technologies for the waste sector, the categorization was carried out regarding specific waste treatment scenarios, whose efficiency is expressed in kg $CO_2$ equivalent emitted per ton of waste. In the USA, with a population of over 300 million people, about 230 million tons per year of the waste are generated which represent about 760 kg per inhabitant per year (OECD). Based on the scenarios discussed above, if all wastes were landfilled, waste disposal would correspond to 425 million tons of $CO_2$ equivalents. Furthermore, Chromec and Ferraro [9] presented the policies of the European Union (EU) on climate and energy. EU has proposed reducing greenhouse gas emissions by at least 40% below 1990 levels. The EU is committed to reducing greenhouse gas *emissions* by at least 40% below 1990 levels by 2030 [9]. In comparison with Europe, annual GHG emissions ($CO_2$-eq/person year) in the USA today are on a level about double that of the Europe. Moreover, in the USA, the EFW concepts are based on the most advanced state of the art, solve a space and pollution problem and guarantee economical and environmentally robust processing and disposal [9].

Chromec and Ferraro [9] determined that if all wastes were incinerated in EFW plants, the emissions could be reduced by about 500 million tons of $CO_2$ equivalents (about 9% of today's USA $CO_2$ output) and make the waste management sector a GHG emission sink. Finally, the total electricity generated from EFW plants could be as high as 15,000 MW replacing about 50 standard 300 MW power plant units [9].

In the other paper, Chromec and Burelle [10] discussed that the maximum environmental benefits from a new EFW facility may require locating the new plant close to both the source of the waste and the potential energy customers. Placing the EFW facility directly into an urban community leads to the following: (*i*) minimizes the cost and the environmental impact of waste transport, (*ii*) allows electrical power to be generated at the point of consumption, (*iii*) provides thermal energy for district heating and cooling, (*iv*) reduces the dependence on imported fossil fuel for electrical generation and for heating/cooling, (*v*) provides secure and well-paying jobs for members of the community and finally reduces the carbon foot print of the community. The two case studies included in this paper described the Isséane EFW plant in Paris and the EFW plant in London [10].

## 2. Example of Incineration plants. Literature review

In Europe, the recovery of EFW has been adopted by the European Commission as one of the sustainable waste management options, with the scope to decrease the amount of nonhazardous waste going to landfill [11]. In the context of waste-to-energy, it is worth noting the European Environmental Citizen Organization for Standardization (ECOS), nonprofit association established in 2002, funded by the European Commission and based in Brussels [12]. ECOS acknowledges that incineration of biomass contribute to climate protection, in addition ECOS admits that energy recovers only if performance is higher (LCA), and waste with low net calorific value or high level of pollutants shall not be used as "fuel" [12].

Waste incineration practices were discussed in the framework of project "QUOVADIS Waste-to-fuel Conversion — A Thinkshop," which took place in Ispra, Italy, between 28 April 2005 and 29 April 2005 [13]. Workshop was the forum for discussion and debate on the exchange of various experiences concerning the use of waste in waste-to-fuel conversion and its subsequent application [13]. In the draft paper untitled "Survey on the on-going scenario for SFR (solid recovery fuels) production and use in Italy" prepared by MP Maranzana presented 44 waste-to energy plants in Italy [13].

With the introduction of legalization, the situation in the EU has been improved markedly, but with pressure to minimize landfill and limited option for recycling, thermal treatment occupies a key role in waste management [14] (ENERGY]. The main challenges to incineration plant operation are maximizing thermal efficiency for energy recovery [14]. The *Clean and Efficient Waste Incineration, Waste-to-Energy and Biomass Combustion* (CLEANWEB) project addresses these challenges. The other project *Performance, Reliability and Emission reduction in Waste INcineration* (PREWIN) Network provides the forum for collaboration among pan-European organizations for waste incineration [14]. One of the objectives of the project is to improve the performance, efficiency and reliability of thermal treatment plant (**Figure 1**).

**Figure 1.**
*Process flow diagram related to CLEANWEB and PREWIN projects [14].*

### 2.1 Maishima incineration plant

Maishima ("dancing island" in Japanese) is a man-made Island in the Bay of Osaka. Osaka is Japan's third largest city, with a population of 2.6 million and with seven completed Von Roll plants [15]. Impressed by a facility in Vienna that had been embellished by the world famous Austrian artist, Osaka's planners decided to entrust him with the design and ornamentation of the building, the stack and the surrounding area (**Figure 2**). The facility consist of two waste processing lines, each of which is able to incinerate 450 tons of MSW every 24 hours, and is typical of today's large scale.

The energy produced in the incineration process is used to generate power in the power generator with a capacity of 32 MW [15].

### 2.2 AVI Moerdijk incineration facility

The Dutch provinces of Zeeland, North Brabant and Limburg funded study in 1990 to determine how much waste was being generated in the southern Netherlands. Based on the 600,000 tons of waste, new waste incineration plant AVI Moerdijk linked to thermal power plant was built. The facility being operation since 1996/1997 and consists of three separate lines. Waste heat generated in a year amounts to 2,000,000 tons of high-pressure steam at temperature of 400°C. The photo and schematic of the AVI Moerdijk incineration facility are presented in **Figures 3** and **4**, respectively [16].

### 2.3 Bergen incineration facility

In the Bergen region on the western coast of Norway, the city Bergen and neighboring communities founded Bergensområdets interkomunale Renovasjonsselskap (BiR) company with the objectives of managing the waste generated by the region's 280,000 inhabitants. The contract for erecting waste incineration plant was awarded to Von Roll Inova in 1997. Start of operation was in 1999. Energy recovery based on 90,000 tons of waste is about 60 GWh of electrical energy and 430,000 tons of process steam. The photo and schematic of the BiR Bergen incineration facility are presented in **Figures 5** and **6**, respectively [17], as well as in **Figure 7** [18].

**Figure 2.**
*Maishima incineration plant of 900 tons/24 hours of waste [15].*

**Figure 3.**
*AVI Moerdijk incineration facility of 600,000 tons/year of waste [16].*

**Figure 4.**
*Scheme of the AVI Moerdijk incineration facility of 600,000 tons/year of waste [16]. 1-Tipping hall, 2-waste pit, 3-overhead crane, 4-crane pulpit, 5-feed hopper, 6-reciprocating grate, 7-primary air supply, 8-secondary air supply, 9-auxilary burners, 10-wet deslagger, 11-slag pit, 12-slag crane, 13-two-pass boiler, 14-boiler ash removal, 15-SNCR DeNOx system, 16-electrostatic precipitator, 17-ash silo, 18-gas-to-gas heat exchanger, 19-wet scrubber, 20-gypsum silo, 21-fabric filter with activated carbon injection, 22-inducted draft fan, 23-stack, 24-emission monitoring.*

### 2.4 KEBAG waste incinerator plant, Zuchwil

According to DOKA [19], Switzerland has 28 municipal solid waste incinerator (MSWI) plants in operation 1 (2000). The majority of them [23] have two or three furnace lines. All Swiss MSWIs utilize the energy contained in the waste to produce useful heat and/or electricity. In 2000, Swiss MSWIs gross production was 2,526,800 MWh heat and 1,284,200 MWh electricity, and the total energy input was 9,880,262 MWh [20]. **Figure 8** displays the general photo of the KEBAG waste incinerator plant in Zuchwil, Switzerland. In its four incineration lines, Zuchwiler KEBAG treats 220,000 tons of flammable domestic waste a year from the cantons of Berne and Solothurn—i.e., from around 473,000 residents in 208 municipalities, as opposed

**Figure 5.**
*BiR Bergen incineration facility of 90,000 tons/year of waste with energy output about 60 GWh. Waste return in form of electricity and heat [17].*

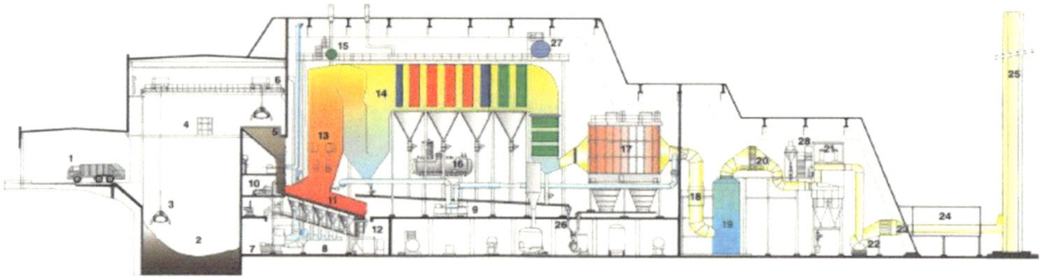

**Figure 6.**
*Scheme of the BiR Bergen incineration facility of 90,000 tons/year of waste [17]. 1-Truck unloading area, 2-waste pit, 3-, 4-crane operating pulpit, 5-feed hopper, 6-primary air intake, 7-primary air fan, 8-primary air distribution, 9-recirculated flue gas fan, 10-ram feeder, 11-reciprocating grate, 12-wet deslagger, 13-auxilary burners, 14-four-pass steam generator, 15-steam drum, 16-feedwater storage tank, 17-electrostatic precipitator, 18-quench, 19-wet scrubber, 20-steam-to-gas heat exchanger, 21-fabric filter, 22-induced draft fan, 23-silencer, 24-emission monitoring, 25-stack, 26-ash removal system, 27-emergency water tank, 28-adsorbent metering station.*

**Figure 7.**
*View general of the BiR Bergen incineration facility (photo from 2018) [18].*

**Figure 8.**
*KEBAG waste incinerator plant, Zuchwil, Switzerland [20]. Note: The name KEBAG comes from the German words for waste disposal.*

to 2002 year, when train 4 enters service, and the service area has around 350,00 habitants who generated some 200,000 tons of wastes. The energy produced in the incineration process is used to generate electricity and prepare hot water [21, 22].

## 2.5 KVA Thun energy-from-waste plant

The KVA Thun energy-from-waste plant handles some 100,000 tons of combustible waste (domestic and bulky) a year. Thun is the city, located on Lake Thun, is the economic hub of the Bernese Mittelland and Oberland with the population of 300,000 residents in 150 communities. The facility produces about a third of the electricity consumed in the city of Thun, as well as provides district heating for different customers facilities (e.g., adjacent public sector facilities [23]). Energy recovery type is extraction-condensation turbine produced electric power of 12 MW (maximum) and district heating output about 25 MW (maximum) [23]. The photo and longitudinal section of the KVA Thun energy-from-waste plant are presented in **Figures 9** and **10**, respectively [22].

The applicable emissions guarantees are below Swiss air quality regulation (LRV) limits, as show in the **Figure 11** below [24].

**Figure 9.**
*KVA Thun energy-from-waste plant of 100,000 tons/year of combustible waste with energy recovery: electric power about 12 MW (maximum) and district heating output about 25 MW (maximum) [23].*

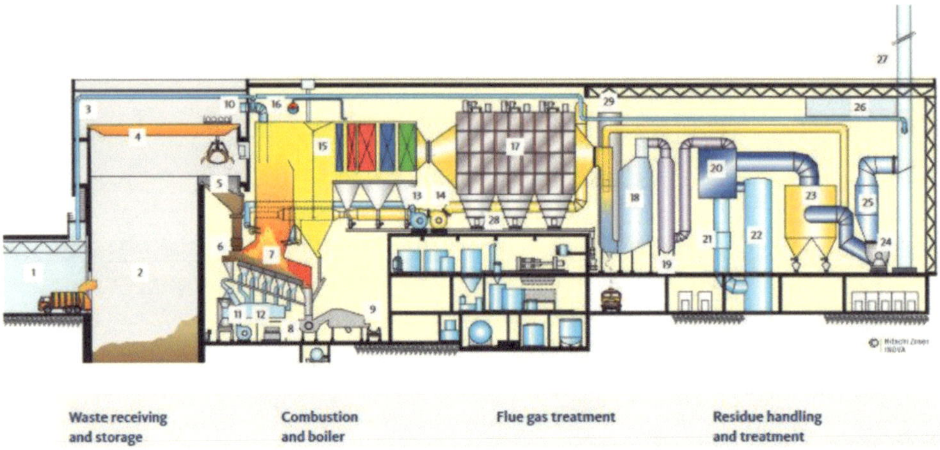


**Figure 10.**
*Longitudinal section of the KVA Thun [23]. 1-Tipping hall, 2-waste bunker, 3-waste pit ventilation, 4-waste crane, 5-feed hopper, 6-ram feeder, 7-hitachi zosen inova grate, 8-ram bottom ash extractor, 9-bottom ash handling, 10-primary air intake, 11-primary air fan, 12-primary air distribution, 13-secondary air fan, 14-recirculation fan, 15-four-pass boiler, 16-boiler, 17-electrostatic precipitator, 18-SCR DeNox with catalyst, 19-economizer, 20-gas/gas heat exchanger, 21-quench, 22-wet scrubber, 23-fabric filter, 24-inducted draft fan, 25-silencer, 26-emissions measurement, 27-stack, 28-ash conveying system, 29-residue silo.*

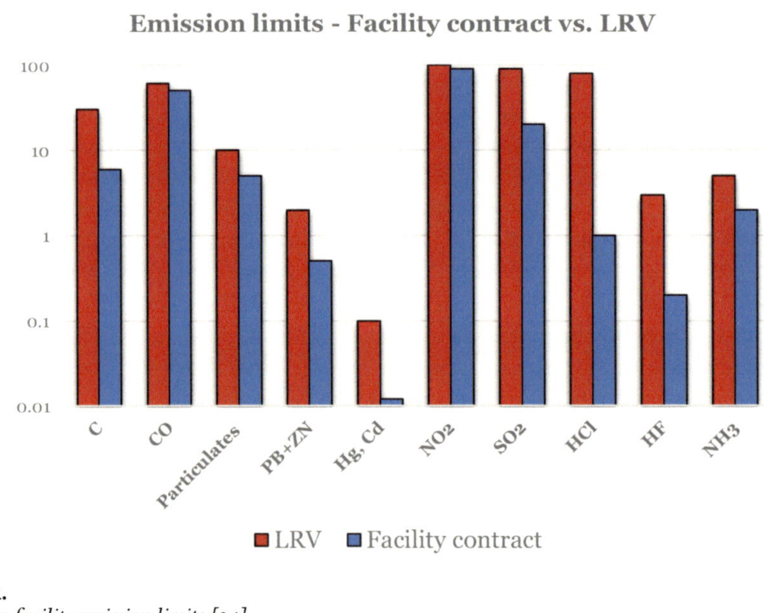


**Figure 11.**
*KVA Thun facility emission limits [24].*

## 3. Forms of recycling in Poland

One recycling method is energy (thermal) recycling, understood as total or partial energy recovery. The waste incineration process must take place under certain conditions due to the toxic substances formed during combustion, such as dioxins or furans emitted and polluting the atmosphere.

The choice of recycling method depends on the type and properties of the post-consumer product. Due to changes in the prices of raw materials and energy, limited possibilities of using material recycling, as well as the fact that raw material recycling can already be considered cost-effective, the role and importance of this form of recycling should increase.

The thermal methods of waste removal that we can distinguish are as follows: combustion, pyrolysis (degassing/gasification), hydrogenation and bios drying. The main waste disposal system is of course incineration, which removes waste that is unusable and cannot be managed further.

Incineration is intended to make the waste residues neutral to the environment and minimize waste gas emissions. It is also important to reduce the volume and use the energy generated during combustion.

The industry must also be guided by several important principles related to the combustion of precipitation:

- operational safety
- investment value
- land demand
- process efficiency and efficiency.

The pyrolysis process breaks down organic waste under the influence of temperature. The product is energy carriers useful for storage. Most combustion technologies are still in the testing phase and are being gradually introduced to the market in mature forms.

Another thermal process to mention is hydrogenation, which uses refinery waste and in some cases plastic waste. However, the goal of the bios drying process is the production of alternative fuel from a biodegradable fraction segregated from mixed municipal waste. The biological drying process is preceded by separation of biodegradable fraction (0–60/80 mm) from mixed municipal waste, followed by mechanical treatment of the dried fraction to produce fuel.

## 4. The thermal waste treatment plant in Krakow (eco-incinerator)

The main objective of this study is to share the art knowledge about the TWTP (eco-incinerator) in Krakow, Poland, that has operated since 2016. Description of TWTP case study is based on the thermal waste treatment plant in Krakow presentations given in [25–28].

In recent years, waste incineration has been frequently preferred to other waste treatment or disposal alternatives due to advantages such as volume reduction, chemical toxicity destruction and energy recovery [29].

According to the study by Pfeiffer [30], energy recovery is a secondary goal of waste incineration: thermal waste treatment and energy recovery are "married" within the waste-to-energy plant [30]. From an economic point of view, a WTE plant treating MSW is an enterprise using a special fuel [1].

Thermal waste treatment plant (TWTP) in Krakow (eco-incinerator) is being constructed in answer to concepts link economic, ecological and social aspects and needs of Krakow as key factor of project "Waste Management Program in Krakow" under the Operational Program Infrastructure and Environment 2007–2013 [27]. On October 31, 2012, a contract was signed with the POSCO Engineering and Construction Co., Ltd., South Korean company, for delivery of a TWTP. On November 6, 2013 began the construction of the eco-incinerator. The contract covered delivery, installation and commissioning the entire electromechanical parts. Series of tests at the plant began from December 3, 2015 until June 27, 2016. According to [26], the total contract amount (net cost) of the project was approximately PLN 666 million (approximately PLN 819 million gross). The subsidy from the European

Union amounted to approximately PLN 372 million (approximately 55.8% of eligible expenses). The contribution of Krakowskiego Holdingu Komunalnego S.A. (KHK) amounted to approximately PLN 294 million and was covered by its own resources and a loan from the National Fund for Environmental Protection and Waste Management (NFEP & WM). The plant is located in the district Nowa Huta, part of the Kraków city. General view and longitudinal section of the TWTP are depicted in **Figures 12** and **13**, respectively [27].

Eco-incinerator allows to process 220,000 tons of municipal waste a year. Selected by the inhabitants mixed municipal solid waste (MSW) and other waste (e.g., resulting from mechanical processing of municipal waste) and following waste recovery processes (i.e., material waste, bulk, rubble) are subject to thermal processing. The wastes are collected only from the municipality of Kraków.

The emissions come from TWTP production process meet the requirements of the best available techniques (BAT), guaranteeing the highest standards of environmental protection.

**Figure 12.**
*The thermal waste treatment plant in Krakow [31].*

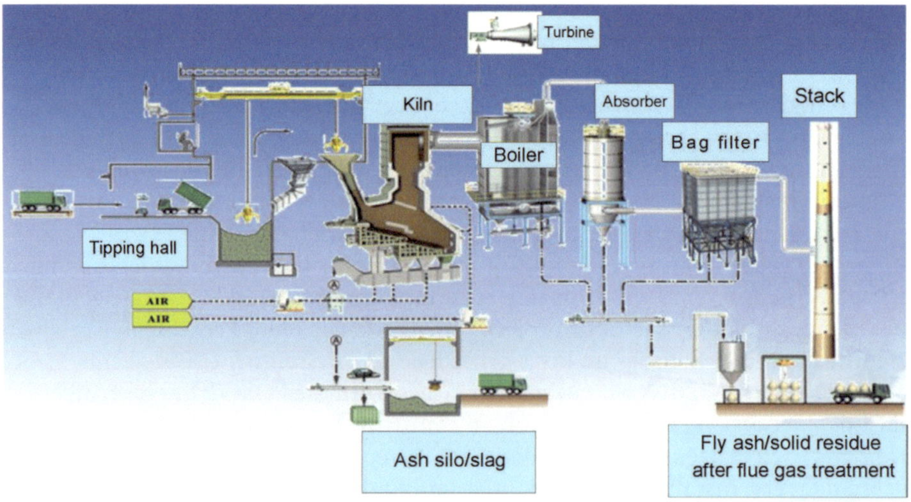


**Figure 13.**
*Longitudinal section of the TWTP plant in Kraków [32].*

On October 1, 2016, the educational pathway, an integral part of the eco-incinerator, was launched in the eco-incinerator area with the goal of increasing the ecological knowledge of the inhabitants as well as bringing the technology and thermal plant operation processes closer to them. This educational pathway is the tour throughout the plant, showing the key steps of the system such as delivery and unloading of waste into the bunker, the process of thermal conversion of municipal waste, the flue gas scrubbing process, the electricity and heat production process and the process of managing residue after incineration [27].

### 4.1 The TWTP-a cost-effective and safe solution

Naturally, very stringent environmental regulations are applied to the process. Using BAT, the TWTP in Krakow meets the highest environmental and legal standards applicable in the European Union. The amount of waste is limited, and thus securing energy to the city has increased.

In the line of [27], energy security signifies a prospective condition of meeting consumer demand (including the citizens of Krakow) for fuel and energy through alternative sources of acquiring gas [27].

### 4.2 Emission standards

According to [28], the plant is equipped with a state-of-the-art exhaust purification system that meets strict emission standards for air.

The exhaust aftertreatment process consists of the following stages:

- Denitrification of exhaust gases
- Flue gas cleaning by means of a semi-dry method
- Dust extraction.

| **Line 1** | **Emission value** | **Emission value** | **Emission value** | **Emission limit value** |
|---|---|---|---|---|
| Date | 27 July 2019 | 13 August 2019 | 20 August 2019 | Standards |
| Dust | 1.070 | 0.990 | 1.090 | 30 |
| Hydrogen chloride | 0.300 | 5.500 | 1.500 | 60 |
| Hydrogen floride | 0.000 | 0.000 | 0.055 | 4 |
| Sulfur dioxide | 1.100 | 19.750 | 12.900 | 200 |
| Carbon monoxide | 12.200 | 7.850 | 6.300 | 100 |
| Nitric oxide | 138.200 | 127.400 | 87.500 | 400 |
| Total organic carbon | 0.300 | 0.300 | 0.600 | 20 |
| **Line 2** | | | | |
| Dust | 1.250 | 1.220 | 1.060 | 30 |
| Hydrogen chloride | 0.100 | 0.500 | 0.300 | 60 |
| Hydrogen floride | 0.000 | 0.000 | 0.020 | 4 |
| Sulfur dioxide | 6.150 | 11.500 | 0.400 | 200 |
| Carbon monoxide | 18.750 | 21.900 | 20.100 | 100 |
| Nitric oxide | 163.400 | 154.600 | 114.100 | 400 |
| Total organic carbon | 0.000 | 0.600 | 0.400 | 20 |

*Concentration magnitude 11% content of oxygen in flue gases (mg/m$^3$), on average in 30 minutes in mg/m$^3$.*

**Table 1.**
*Emission values of the TWTP derived from measurements of 27 July and 13 August 2019 [28].*

The emission standards comply with the Ordinance of the Minister of the Environment of 4 November 2014 on emission standards for certain types of installations, sources of fuel combustion and devices for incineration or co-incineration of waste (Journal of Laws of 2014, item 1546, including further amendments).

Below you can find in **Table 1** [28] a sample of selected emission values derived from measurements of 27 July, 13 August and 20 August this year.

As mentioned above, the eco-incinerator was created as a response to the ecological and energy needs of Krakow as part of the project "Municipal Waste Management Program in Krakow" (TWTP). It is the latest and most important part of this system; it enables to utilize municipal waste generated by the inhabitants of Krakow and the recovery of energy from it.

The thermal recycling technology is the most mature and environmentally esponsible solution to waste. This is confirmed by many years of European experience in which thermal processing of waste with recovery of energy forms the basis of the entire waste management system.

The eco-incinerator allows to process 220,000 tons of municipal waste during the year. Approximately 65,000 MWh of electricity and 280,000 MWh of heat are produced as a result of the combustion. The energy obtained by way of the thermal transformation process is largely organic and renewable [27].

### 4.3 Green energy factory

The thermal waste treatment plant in Krakow called Green-energy factory [26] provides background material for this part.

Concept of the furnace-grate furnace integrated into the boiler.

- Grate incinerator technology
- Boiler—drum-type heat recovery steam generator (HRSG) with natural circulation
- Turbine—based on the extraction condensing process [26].

### 4.4 The TWTP receives

- Municipal solid waste mixed with secondary raw materials separated.
- Combustible ballast from other installations for processing municipal [26].

### 4.5 Selection of waste, unloading and loading in the combustion chamber

The overall objectives of the TWTP are to increase thermal efficiency and heat recovery, as well as reduce emissions through improving and developing a better plant characteristics. For this purpose, the wastes are discharged in the unloading hall of the bunker that ensures stocking waste for 5 days. To obtain a uniform mixture, which improves incineration efficiency, the waste is mixed together in the bunker. Next the waste is mixed together in the bunker, and then the cranes equipped with 6-teeth grapple buckets transport mixed waste to feed hoppers where it feeds into the loading shaft [25].

In the TWTP were installed radioactive material detectors so as to protect the installations against damage. The TWTP also monitors waste delivered for thermal processing [26].

### 4.6 Facility (TWTP) concept

The concept of the thermal waste treatment process is based on the five following steps:

#### *4.6.1 Drying process*

Radiation or convection is used for the waste drying in the grate zone. Thanks to temperature at approximately 100°C the moisture through evaporation is eliminated [26].

#### *4.6.2 Degassing process*

The waste continued heating at above 250°C is used to releases gases (e.g., moisture and carbonization gases) [26].

#### *4.6.3 Combustion process*

Third grate zone is suitable for waste incineration where loss on ignition is below 0.5% of the mass share production [26].

#### *4.6.4 Gasification process*

This process achieves oxidation volatile products by molecular oxygen. The majority of the waste is oxidized at 1000°C in the upper section of the furnace chamber [26].

#### *4.6.5 After burning*

The carbon dioxide amount in the combustion gases is reduced in the recuperative thermal oxidation area. The standard operating procedure used secondary air for total incineration for the minimum at 850°C and trough minimum of 2 seconds [26].

The chosen combustion technology provides in the combustion chamber the reduction of CO and $NO_x$, dioxins and furans emission. Produced in this thermal waste processing energy is then optimally recovered in a heat recovery steam generator (HRSG) integrated with the grate furnace [26].

### 4.7 Energy recovery process

Using innovative concept, the heat recovery steam generator (HRSG) is based on natural circulation of exhaust gases. In the boiler after heat exchange, exhaust gases are cooled to approximately 180°C. In the next step, heat is used to convert water, via the boiler, into superheated steam. Superheated steam at a pressure of 40 bar and a temperature of 415°C is supplied to the electricity generation and transmission node. The produced electricity is used to drive a set of steam turbines. In the last step, the actuated turbine converts mechanical energy into electrical energy [26].

According to [26], produced electricity is used (consumed) by the operator; on the other hand, energy excess is returned to the grid. Using proven cogeneration concept, produced heat is distributed to the municipal heating network [26].

### 4.8 Emission treatment process

In line with [26], carbon dioxide, carbon monoxide, sulfur dioxide, oxides of nitrogen, steam and unburned or partially burned hydrocarbons are generated during the waste combustion process. Pollution may be expressed both in the form of gas and dust [26].

Pollution control plays an important role in waste incineration. All existing emission limits are complied. Combustion gases pass first to the heat recovery steam generator. Next via emission treatment installation are transferred to gas exhaust fan and finally are removed to stack.

Procedure of the emission treatment consists of the following steps [26]:

- Exhaust gases denitrification used by primary and secondary methods based on the application of the selective nitrogen oxides reduction (SNCR) via the injection of aqueous urea solution containing 25% urea by weight.

- Semi-dry method based on the use of lime slurry, together with the dry entrained flow method with activated carbon. These methods reduced acidic impurities, as well as heavy metals, dust, furans and dioxins.

- Fabric filtration for the effective dedusting exhaust gases by means of fabric filtration [26].

Post-processing waste disposal:

According to the [26], by-products created during thermal processing of waste are as follows:

- slag and bottom ash

- boiler dust

- fly ash

- solid residues from the emission treatment.

Each residue resulted from thermal waste processing shall be received by authorized external entities in accordance with the provisions of the IPPC standards [26].

## 5. Conclusions

1. The use of waste as a fuel leads to saving of primary resources and the reduction of $CO_2$ emissions.

2. From an economics point of view, it is clear that waste incineration is costly for society.

3. The use of waste as alternative fuels is carried out in total compliance with permits issues by relevant authorities and meets rigorous requirements according to EU and national law and regulations as well as according to the IPPC Directive.

4. The main advantage that gasification has over incineration is its capability to retain (keep) the chemical energy of the waste in the produced syngas.

## Acknowledgements

Relevant figures are reprinted with given sources.

## Conflict of interest

The authors declare that they have no conflict of interest. The research does not involve human participants and/or animals.

## Author details

Dariusz Sala* and Bogusław Bieda
AGH University of Science and Technology, Management Department, Kraków, Poland

*Address all correspondence to: dsala@zarz.agh.edu.pl

## References


[1] Bieda B. The role of thermal treatment in an integrated waste management. Waste recycling. In: Wzorek Z, Kulczycka J, Fečko P, Kušnierowa M, editors. Kraków, Poland: Mineral and Energy Economy Research Institute of the Polish Academy of Sciences. 2005. pp. 104-113

[2] Everard M. Hot Stuff!–The Role of Thermal Treatment in a Sustainable Society. Warlines Park House, Horseshoe Hill, Upshire, Essex, UK: Waste Management World, PennWell International Publications Ltd.; 2004. pp. 37-45

[3] Hulgaard T, Vehlow J. Incineration: Process and technology. In: Christensen TH, editor. Solid Waste Technology & Management. Vol. 1. Chichester: Blackwell Publishing Ltd.; 2011. pp. 365-392

[4] Moberg A, Finnveden G, Johansson J, Lind P. Life cycle assessment of energy from solid waste–Part 2: Landfilling compared to other treatment methods. Journal of Cleaner Production. 2015;**13**(3):231-240

[5] Eriksson O, Finnveden G, Ekvall T, Bjorklund A. Life cycle assessment of fuels for district heating: A comparison of waste incineration, biomass- and natural gas combustion. Energy Policy. 2007;**35**(2):1346-1362

[6] Hauck P, Strobridge D, Sonawane A. Sustainable Solutions for the 21st Century. Santa Barbara, CA: MSW Management; Forester Media Inc.; 2012. pp. 12-21

[7] Chandler AJ, Eighmy TT, Hartlén J, Hjelmar O, Kosson DS, Sawell SE, et al. Municipal Solid Waste Incinerator Residues. Studies in Environmental Science 67. Amsterdam, The Netherlands: The International Ash Working Group. Elsevier Science; 1997. p. 973

[8] Kleis H, Dalager S. 100 Years of Waste Incineration in Denmark. From Refuse Destruction Plants to High-Technology Energy Works. Copenhagen, Denmark: Babcock and Wilcox, Vølund and Rambøll; 2004. 51p

[9] Chromec PR, Ferraro AF. Waste-to-energy in the context of global warming. In: 16th Annual North American Waste-to-Energy Conference. 2008. DOI: 10.1115/NAWTEC16-1954

[10] Chromec PR, Burelle RJ. Integration of an energy from Waste facility into an urban environment. In: 17th Annual North American Waste-to-Energy Conference. 2009. DOI: 10.1115/NAWTEC17-2320

[11] Ciceri G. QUOVADIS-quality management organisation, validation of standards, developments and inquiries for SRF. QUOVADIS waste-to-fuel conversion?–A Thinkshop. DG JRC Workshop, Ispra, 28-29 April 2005. Workshop Proceedings. In: Gawlik BM, Ciceri G, editors. EUR Report 21756 EN. European Communities. Luxembourg: Office for Official Communications of the European Communities; 2005. pp. 147-155

[12] Tebert C. Waste-to-fuel standarisation. The public confidence aspect. QUOVADIS Waste-to-fuel conversion?–A Thinkshop. DG JRC Workshop, Ispra, 28-29 April 2005. Workshop Proceedings. In: Gawlik BM, Ciceri G, editors. EUR Report 21756 EN. European Communities. Luxembourg: Office for Official Communications of the European Communities; 2005. pp. 123-128

[13] Maranzana M. Survey on the on-going scenario for SFR (solid recovery fuels) production and use in Italy. In: Gawlik BM, Ciceri G, editors. QUOVADIS waste-to-fuel conversion?–A Thinkshop. DG JRC

Workshop, Ispra, 28-29 April 2005. Workshop Proceedings. In: EUR Report 21756 EN. European Communities. Luxembourg: Office for Official Communications of the European Communities; 2005. pp. 209-232

[14] McGarry D. Energy from Waste and Biomass for better Efficiency and a Cleaner Environment. European Commision-Directorate General, Joint Research Centre, Institute for Energy. Petten, NL: P.O. Public Relations and Communication; 2005. p. 2

[15] Ermel FW. Maishima-The Latest Technology and a Unique Design. Flas VonRoll INOVA. nr 21. Zurich, Switzerland: Von Roll Technology Environmental Ltd; 2003. 5p

[16] AVI Moerdijk. vonRoll INOVA's Leaflet. Von Roll Environnement SA. Hardturmstrasse 133, Case postale 760, CH-8037 Zurich. Available from: http://www.vonroll.ch/inova [Accessed: 08 August 2019]

[17] BiR Bergen. vonRoll INOVA's Leaflet. Von Roll Environnement SA. Hardturmstrasse 133, Case postale 760, CH-8037 Zurich Available from: http://www.vonroll.ch/inova [Accessed: 08 August 2019]

[18] Ristesund A. Bellona jubler for BIR-initiativ. Foto: Forbrenninganlegget til BIR i Rådalen. 2018. Available from: https://www.ba.no/nyheter/miljo/bir/bellona-jubler-for-bir-initiativ/s/5-8-852769 [Accessed: 06 December 2019]

[19] Doka G. Part II Waste Incineration. Life Cycle Inventories of Waste Treatment Services.ecoinvent report No. 13, Swiss Centre for Life Cycle Inventories, Dübendorf, December 2003. Available from: https://www.doka.ch/13_II_WasteIncineration.pdf [Accessed: 10 August 2019]

[20] BUWAL. Energieerzeugung und Nutzung in KVA 2000, Stand 11.12.2001. Industrie- und Gewerbeabfälle, Berne, Switzerland. 2001. Available from: http://www.buwal.ch/abfall/docu/pdf/tab10_d.pdf [Accessed: 10 August 2019]

[21] Contract successfully executed for KEBAG Zuchwil. 2012. Available from: https://www.alpiq.com/alpiq-group/media-relations/news-stories/news-stories-detail/contract-successfully-executed-for-kebag-zuchwil/ and https://www.alpiq.com/search/search-results/?L=0&id=119&tx_solr%5Bq%5D=zuchwil [Accessed: 08 August 2019]

[22] KEBAG Waste Incinerator Plant, Zuchwil, Schweitz. Available from: https://nightnurse.ch/en/portfolio/project/kebag-waste-incinerator-plant-zuchwil-schweiz-277 [Accessed: 10 August 2019]

[23] Thun/Switzerland. Hitachi Zosen Inova. Energy-from-Waste Plant. Available from: http://www.hz-inova.com/cms/wp-content/uploads/2014/11/hzi_ref_thun_en.pdf [Accessed: 10 August 2019]

[24] Surber M. Thun Waste Treatment Facility. Flas VonRoll INOVA. Nr 20. Zurich, Switzerland: Von Roll Technology Environmental Ltd; 2002. 5p

[25] The Thermal Waste Treatment Plant in Krakow. ECO-INCINERATOR. Available from: https://khk.krakow.pl/en/eco-incinerator/ [Accessed: 08 August 2019]

[26] The Thermal Waste Treatment Plant in Krakow. Green Energy Factory. Available from: https://khk.krakow.pl/en/eco-incinerator/green-energy-factory/ [Accessed: 08 August 2019]

[27] The Thermal Waste Treatment Plant in Krakow. Available from: https://khk.krakow.pl/en/eco-incinerator/about-the-plant/ [Accessed: 08 August 2019]

[28] The Thermal Waste Treatment Plant in Krakow. Available from: https://khk.krakow.pl/en/eco-incinerator/emission/ [Accessed: 10 August 2019]

[29] Sonnemann G, Castells F, Schumacher M. Integrated Life-Cycle and Risk Assessment for Industrial Processes. Boca Raton, FL, USA: CRC Press Company, Lewis Publishers; 2004. 392p

[30] Pfeiffer E. Waste or Valuable Product? Warlines Park House, Horseshoe Hill, Upshire, Essex, UK: Waste Management World, PennWell International Publications Ltd.; 2004. pp. 65-69

[31] Eco-Incinerator. Photos. Available from: https://www.google.pl/search?q=ekospalarnia+krak%C3%B3w&tbm=isch&source=iu&ictx=1&fir=ar1943G_9-iQCM%253A%252ChQN9Zq2XdEMgsM%252C_&vet=1&usg=AI4_-kQEzy-yzl0DkU7Xv8e9_8A9ABJa9Q&sa=X&ved=2ahUKEwis2tHQvJPkAhVRkMMKHVTsD54Q9QEwEnoECAUQBg#imgrc=sX8Vh7uDTuZQnM:&vet=1 [Accessed: 18 August 2019]

[32] POSCO Project Presentation. The Thermal Waste Treatment Plant Conference in Kraków. Public Presentation. Kraków, Poland. 3 December 2013

Chapter 7

# Management of Coal Fly Ash in Remediation Process

*Markandeya, S.P. Shukla and Devendra Mohan*

## Abstract

The present research relates to class of adsorbents obtained by systematic biopolymer modification of cenospheres transfigured from coal fly ash (CFA): an immense waste by-product of coal based thermal power plant, method of preparation thereof and their use in wastewater treatment contaminated by tanneries, distilleries, cosmetics, textiles, plastics, pulp and paper industries, paints, electroplating and food processing industries effluents. Removal percentage of disperse dyes had better correlated with Langmuir isotherm, tested among Freundlich, Temkin and Redlich-Peterson isotherm which indicated saturated monolayer attachment of dye molecules onto the surface of adsorbent with maximum capacity 500.4 and 500.0 mg/g for Disperse Orange 25 (DO) and Disperse Blue 79:1 (DB) dyes, respectively. The uptake rate of dye molecules followed pseudo-second order kinetics in all cases. Recovery of dye molecules was completed best in three cycles with acetic acid for CFA and cenospheres, with Di-chloromethane for CNAC and in four cycles with non-polar solvent (chloroform) for zeolite and CNCH nanocomposite. The used adsorbents could easily be dumped into landfill with in concrete pit liming, or can also be used in brick making to minimize the environmental risk.

**Keywords:** cenospheres, coal fly ash, disperse dyes, response surface methodology, waste management

## 1. Introduction

Management of coal fly ash (CFA) at 145 coal based thermal power plants in India is a difficult task as large quantity of ash being generated. MoEF & CC had fixed the target of 100% utilization of CFA in time bound manner [1]. The average utilization percentage of CFA in various fields is about 62% only. Remaining 90 million ton of CFA is disposed into holding ponds, lagoons, landfills and slag heaps which not only requires large area of precious land for its disposal but is also one of the sources of pollution of both air and water. The utilization of CFA in 1990s was 40 million ton (3% utilization) annually and about 147 million ton (62% utilization) in 2015s [2]. The land for creating ash dyke for ash disposal facilities at thermal power plants is becoming difficult to be acquired. Thereby, CFA, if not managed well, may pose various environmental challenges.



IntechOpen

## 2. Methodology

Characterization and batch adsorption study with four modified CFA adsorbents has been presented in details (**Figure 1**).

Legislative council passed a law to eliminate color from their industrial effluent before discharging dye-containing effluents into water bodies because dyes have synthetic origin and fused complex aromatic structure making them more recalcitrant to biodegradation [3, 4] (**Figure 2**).

Adsorption has evolved into one of the most effective and feasible physical processes for decolorization of textile and dyeing wastewater [5–7]. Though there is no dearth of available adsorbents like activated carbon, economic factors force consideration of alternative low cost, eco-friendly and efficient adsorbents which are either naturally available or available as waste products from other manufacturing processes and which can be utilized for the treatment of industrial effluents by the entrepreneurs of small scale industries in developing countries like India [8, 9].

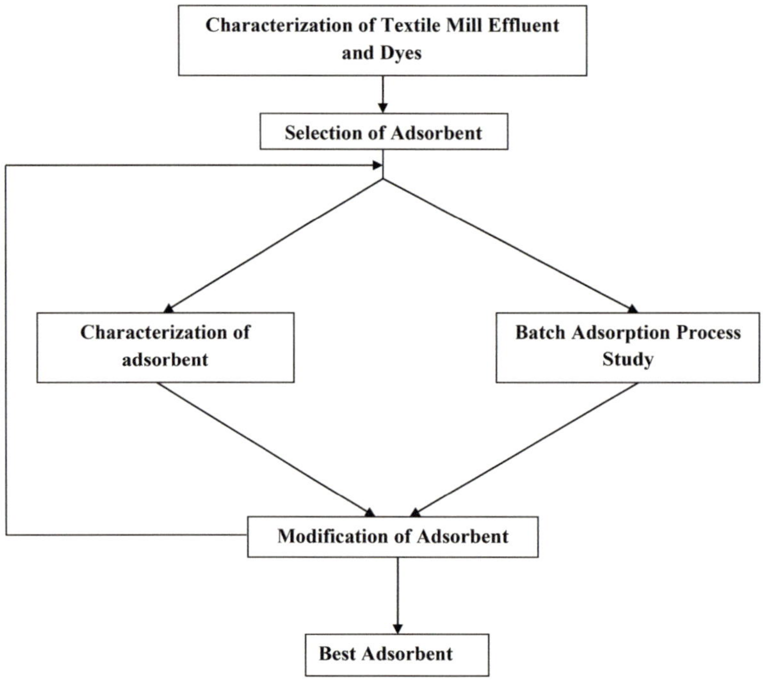


**Figure 1.**
*General layout of study design.*

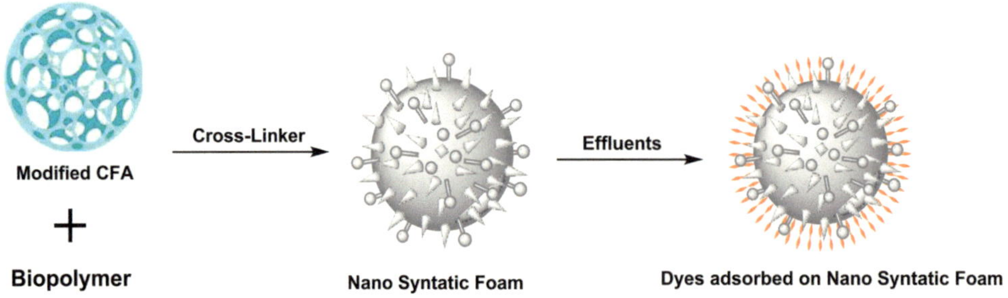


**Figure 2.**
*Process of dye adsorption.*

In this backdrop coal fly ash (CFA) has been actively pursued for its decolorizing characteristics on textile mill effluents but its relatively lower removal efficiency necessitates further modifications. Here the characterization and modification of CFA has been done to obtain the best adsorbent and the optimal conditions in which its removal efficiency for disperse dyes is maximum. To prove its efficiency, characterization of activated carbon and its comparison with our best found adsorbent have also been done [10].

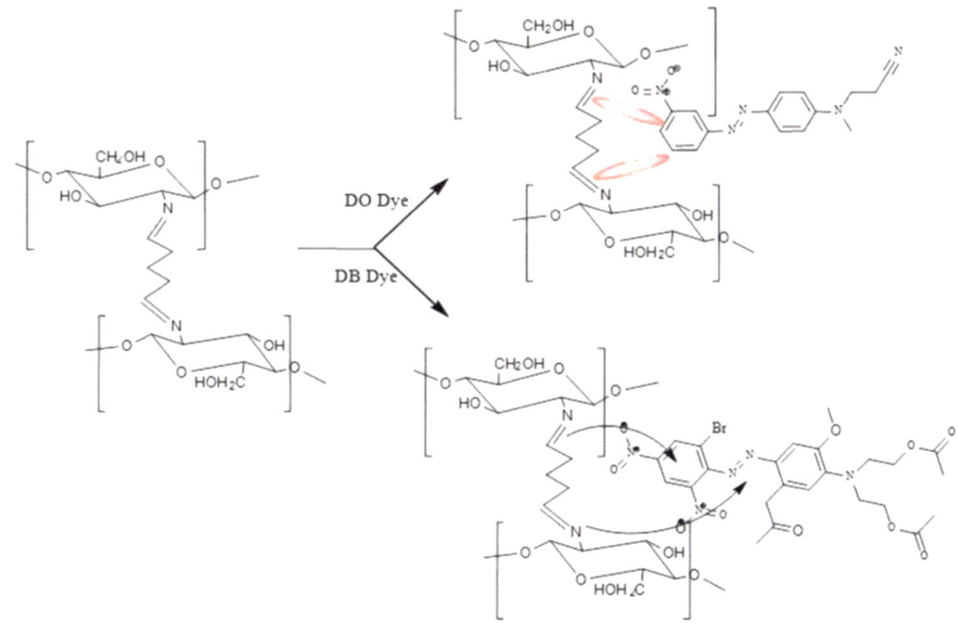


**Figure 3.**
*Interaction between dye molecules and cross-linked natural polymer syntactic nano-foam.*

## 3. Results and discussion

The major elemental leaching of CFA was found to be Fe > Mn > Mg. In the successive stages of research, CFA was modified and characterization studies including ATR-FTIR, SEM, EDX, XRD, TEM, CILAS and BET were conducted to analyze the adsorption potential of each modified form on the basis of its removal efficiency studied for both DO and DB dyes. From the obtained results it could be concluded that CFA has the potential to adsorb Disperse Orange 25 (DO) and Disperse Blue 79:1 (DB) dyes with 75 and 71% removal efficiency at optimized conditions and this percentage removal increased with each modification [81 and 78% in case of cenospheres, 79 and 75% for cenospheres activated carbon (CNAC) composite, 93 and 88% for zeolite, 90 and 87% for cenospheres chitosan (CNCH) nanocomposite], respectively [10–12] (**Figure 3**).

More particularly, the synthesized natural polymer syntactic nano-foam adsorbent displayed high adsorption capacities towards different class of organic dyes (hydrophobic and hydrophilic) via π-π or electrostatic interaction depending upon the nature of the dye molecules (**Table 1**).

In each case, it was found that the percentage removal of DO/DB dye had good vibes with Langmuir isotherm, which indicated saturated monolayer attachment of dye molecules onto the adsorbent with maximum capacity 1.70 and 1.55 mg/g in case of CFA, 33.33 and 32.26 mg/g in case of cenospheres, 90.91 and 83.33 mg/g

| Parameters | CFA | | Cenospheres | | CNAC | | Zeolite | | CNCH | | AC | |
|---|---|---|---|---|---|---|---|---|---|---|---|---|
| | DO | DB | DO | DB | DO | DB | DO | DB | DO | DB | DO | DB |
| pH | 6.0 | | 6.0 | | 6.0 | | 6.0 | | 6.0 | | 6.0 | |
| Adsorbent dosage (g/L) | 60 | 80 | 2.0 | 3.0 | 0.3 AC + 1.0 CN | 0.4 AC + 1.0 CN | 2.0 | 3.0 | 0.2 | | 0.4 | 0.7 |
| Adsorbate conc. (mg/L) | 40 | | 40 | | 40 | | 40 | | 40 | | 40 | |
| Agitation speed (rpm) | 140 | | 140 | | 140 | | 140 | | 200 | | 140 | |
| Cont. time (min) | 100 | 120 | 100 | 120 | 100 | 120 | 100 | 120 | 90 | 120 | 100 | 120 |
| % Removal | 74 | 71 | 80 | 79 | 79 | 75 | 93 | 89 | 90 | 87 | 81 | 77 |

**Table 1.**
*Effect of various parameters for the removal of dyes on various forms of adsorbents.*

| Adsorbents | DO dye (%) | DB dye (%) | Solvents |
|---|---|---|---|
| Coal fly ash (CFA) | 46 | 39 | Acetic acid |
| Cenospheres | 48 | 43 | Acetic acid |
| Cenospheres activated (CNAC) composite | 59 | 54 | Di-chloromethane |
| Zeolite | 46 | 41 | Chloroform |
| Cenospheres chitosan (CNCH) nanocomposite | 47 | 41 | Chloroform |
| Activated carbon | 48 | 41 | Acetic acid |

**Table 2.**
*Recovery and reusability of various adsorbents.*

in case of CNAC, 125.00 and 109.80 mg/g in case of zeolite and 500.40 and 500.00 mg/g in case of CNCH nanocomposite for DO/DB dyes, respectively. The adsorption of dye molecules followed pseudo-second order kinetics in all cases. Thermodynamic study showed that process of adsorption is exothermic in nature. Recovery of dyes was completed best in three cycles with acetic acid for CFA and cenospheres, with Di-chloromethane for CNAC and in four cycles with non-polar solvent (chloroform) for zeolite and CNCH nanocomposite (**Table 2**).

After completing the analysis on the removal efficiency of each modified form of CFA, in the next stage of this work, optimization studies were done. The batch optimization process focused on effect of operating variables, such as contact time, pH, agitation speed, adsorbent dose and dye concentration using RSM with BBD. Furthermore, interactions among various parameters were investigated by applying RSM. In case of zeolite, related $R^2$ values of 0.9102 and 0.9038 were obtained for DO and DB dyes respectively. Optimum results indicated that 119 min of contact time was required to achieve 96% of DO dye removal at zeolite dose 0.67 g/L, dye concentration of 38 mg/L and shaking speed of 158 rpm at pH 6.10. Whereas, 95.23% of DB dye removal was found at contact time of 122 min, adsorbent dose of 1.05 g/L, dye concentration of 26.72 mg/L, agitation speed of 145 rpm and pH of 5.68. Regression modeling and ANOVA showed contact time, adsorbent dose, dye concentration, agitation speed and pH have values of 'Prob. >F'< 0.0500, which indicated that model terms were significant for adsorption of the dyes. For CNCH nanocomposites, F-values 19.91 and 19.26 for DO and DB dyes have a low probability value (Prob. >F < 0.0001), which indicate model terms are symbolic. The high $R^2$ value for DO (0.9409) and DB (0.9391) also showed significance of applied model. The maximum percentage removal of dyes was found to be 97.30% (DO) and 94.22% (DB). From the optimization studies, it could safely be concluded that both zeolite and CNCH nanocomposite had high potential as an efficient, eco-friendly and economic adsorbent for dye removal from aqueous solutions.

## 4. Conclusion

Comparing the results of the characterization and optimization studies, it could be concluded that CNCH nanocomposite was a better adsorbent than all the other modified and native form of CFA. To validate the result, CNCH nanocomposite was compared with the most widely used commercial adsorbent i.e., activated carbon. Results concluded that adsorption capacity of our best found adsorbent was more than two times of activated carbon. To further vindicate the claim that CNCH nanocomposite had high potential as an efficient, eco-friendly and economic adsorbent, it was used for the color removal from textile mill effluent. The adsorbent showed excellent removal efficiency and it was found that physico-chemical parameters also reduced after treatment of effluent. The effect of pH revealed that all the adsorbents produced good results at pH 6 implying that there was no requirement of any special acidic or basic chemical for dye removal. This pH 6 also falls within the limit of pH 5.5–9.0, i.e., industrial effluent discharge limit into the inland surface water as mandated by CPCB. Results also suggest that above CFA based adsorbents could be suitably used as alternative and effective resource materials as compared to the expensive commercial adsorbents for the removal of DO and DB dyes from colored effluents. The used adsorbents could easily be dumped into landfill with liming, in concrete pits or can also be used in brick making to minimize the environmental risk. The present work is also a forward step in reducing the pollution load by utilization of waste products in an eco-friendly manner to effectuate the goal of Agenda 21 and the Rio Declaration on Environment and Development.

## Acknowledgements

The first author is highly grateful to Ms. Vibhuti Mishra for her valuable suggestions.

## Conflict of interest

On behalf of all authors, the corresponding author states that there is no conflict of interest.

## Author details

Markandeya[1*], S.P. Shukla[2,3] and Devendra Mohan[1]

1 Department of Civil Engineering, Indian Institute of Technology-BHU, Varanasi, India

2 Department of Civil Engineering, Institute of Engineering and Technology, Lucknow, India

3 Rajkiya Engineering College, Banda, India

*Address all correspondence to: mktiwariiet@gmail.com; markandeya.civ@iitbhu.ac.in

## References

[1] Government of India, Ministry of Environment and Forests. Fly Ash Notification. 2009. Available from: http://dste.puducherry.gov.in/Flyash_notification2009.pdf

[2] Central Electricity Authority of India. Annual Report on Fly Ash Utilization. 2015. Available from: cea.nic.in/reports/others/thermal/tcd/flyash_final_1415.pdf

[3] Dhiman N, Markandeya, Fatima F, Roy S, Rout PK, Saxena PN, et al. Predictive modeling and validation of arsenite removal by one pot synthesized bioceramic buttressed manganese doped iron oxide nanoplatform. RSC: Advances. 2017;**7**:32866-32876

[4] Dhiman N, Markandeya, Singh A, Verma NK, Ajaria N, Patnaik S. Statistical optimization and artificial neural network modeling for acridine orange dye degradation using in-situ synthesized polymer capped ZnO nanoparticles. Journal of Colloid and Interface Science. 2017;**493**:295-306

[5] Markandeya, Dhiman N, Shukla SP, Kisku GC. Statistical optimization of process parameters for removal of dyes from wastewater on chitosan cenospheres nanocomposite using response surface methodology. Journal of Cleaner Production. 2017;**149**:597-606

[6] Markandeya, Shukla SP, Dhiman N. Characterization and adsorption of disperse dyes from wastewater onto cenospheres activated carbon composites. Environmental Earth Sciences. 2017;**76**:702-714

[7] Shukla SP, Sonam, Markandeya, Mohan D, Pandey G. Removal of fluoride from aqueous solution using *Psidium guajava* leaves. Desalination and Water Treatment. 2017;**62**:418-425

[8] Markandeya, Dhiman N, Shukla SP, Mohan D, Kisku GC, Patnaik S. Comprehensive remediation study of disperse dyes containing wastewater by using environmental benign, low cost cenospheres nanosyntactic foam. Journal of Cleaner Production. 2018;**182**:206-216

[9] Tiwari M, Shukla SP, Mohan D, Bhargava DS, Kisku GC. Modified cenospheres as an adsorbent for the removal of disperse dyes. Advances in Environmental Chemistry. 2015;**2015**:1-8

[10] Kisku GC, Markandeya, Shukla SP, Singh DS, Murthy RC. Characterization and adsorptive capacity of coal fly ash from aqueous solutions of disperse blue and disperse orange dyes. Environmental Earth Sciences. 2015;**74**(2):1125-1135

[11] Markandeya, Shukla SP, Dhiman N, Mohan D, Kisku GC, Roy S. An efficient removal of disperse dye from wastewater using zeolite synthesized from cenospheres. Journal of Hazardous, Toxic, and Radioactive Waste. 2017;**21**(4):04017017

[12] Markandeya, Shukla SP, Mohan D. Toxicity of disperse dyes and its removal from wastewater using various adsorbents: A review. Research Journal of Environmental Toxicology. 2017;**9**:01-18

Section 4

# Global Green Technologies – Economics and Innovation

Chapter 8

# Circular Economy: The Strategies to Global Business Economics

*Woodrow W. Clark II and Danilo Bonato*

## Abstract

Economics needs to change from being a linear theory into being circular because of the need to be more scientific due to the use of both quantitative and qualitative, which reflects science in theory and practice. The application of circular economics to businesses, both in existing and new entrepreneurship, provides a construct way to address, change and reverse climate problems.

**Keywords:** climate, qualitative, quantitative, economics, science, theory, practice

## 1. Overview

Europe and the whole world are facing several pressing issues concerning climate change, the environment, society and the economy, which are crucial to the quality of life. The **Circular Economy** (CE) is the answer to some of the main challenges of our time. It helps today to preserve resources that are increasingly scarce and subject to greater than ever, environmental pressure. It boosts Europe's economy and competitiveness, by generating new business opportunities as well as innovative and more efficient ways of producing and consuming. It brings local jobs and creates opportunities for social integration and cohesion. And even finds an answer to the terror of the fanatics: provide desperate people with viable, safe and strong future for their families and children.

## 2. The European Union: Horizon 2020

For these reasons, in 2013 the European Commission launched a new and comprehensive research and innovation program called Horizon 2020 which will have to introduce a more focused approach to concrete solutions to climate change and environmental challenges. The European Union Circular Economy is a long-awaited package that will play a key role in supporting this transition to the future, today, by providing a clear message to the industry and society on the pathway forward. The package drives investments and creates a level playing field, removing obstacles stemming from European legislation, deepening the single market, and providing favorable conditions for innovation. Experts agree that it is of particular importance that this program is not only convincing from the point of view of research, but that it is also relevant in terms of contribution to the achievement of the objectives of the European Union, including prosperity, quality of life, sustainability, growth and employment.

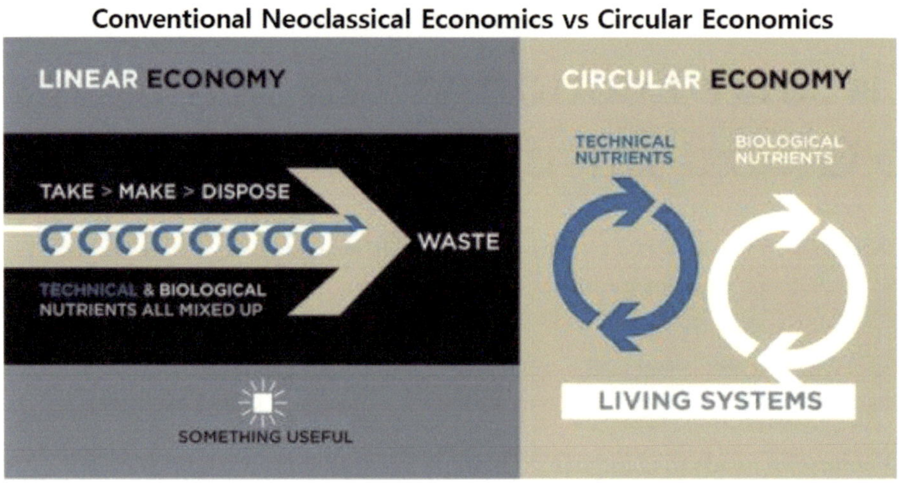


**Figure 1.**
*Danilo Bonato.*

One of the main objectives of Horizon 2020 is to position Europe as leader in the development of a "circular" and "green" economy, based on the concept of sustainability: at this particular moment in time the EU must strengthen its international competitiveness through the productivity of resources and the improvement of the capacity to provide the world with low environmental impact technologies and services, oriented towards an efficient use of resources.

This is therefore an epochal change that must move us away from the conventional neo-classical economy of Adam Smith as Clark and Fast documented and established in their first book, Qualitative Economics [1, 2] that includes other aspects in our development models as well as the "market".

The chart below shows how the Circular Economy works, and has been working in different countries and communities already (**Figure 1**).

The European Union (and the world) is facing a series of systemic crises affecting the environment, natural resources, economic and social aspects, with very critical implications such as population growth, conflicts and wars. To address these systemic challenges successfully, we need eco-innovative solutions, capable of transforming economies and making the lifestyle of European citizens more sustainable as Clark and Cook documented in their book on The Green Industrial Revolution (English 2015 and Mandarin 2014) as well as others about the technological, science, public policy and economic solutions to climate change.

## 3. China: green development

The "disruptive green revolution" (GIR) as "green development" in China has already begun in Europe and is now penetrating China and other parts of Asia.

For example, on 16 July 2018, China and the EU signed a MOU on working together with Circular Economics. Then in late fall 2018, a report was done and released for China to plan and work in this area as a key part of the Chinese Plan for Green Development and a New Silk road starting from east coast of China [3] (**Figure 2**).

We need a solid systemic eco-innovation, inserted in a broader perspective that considers all the innovative solutions able to contribute to the development of the three fundamental "axes" of sustainability: economic, social and environmental.

*DOI: http://dx.doi.org/10.5772/intechopen.88617*

**Figure 2.**
*Wikipedia: Shanghai, China.*

From this point of view, the circular economy is the most effective approach in terms of the model of development and has recently become the strategic reference areas for ALL the EU countries and also thanks to the commitment of the Belgian Presidency of the Council of the Union European Union (Flemish Environment Minister Joke Schauvliege).

MEMORANDUM OF UNDERSTANDING ON CIRCULAR ECONOMY BETWEEN THE EUROPEAN COMMISSION AND THE NATIONAL DEVELOPMENT AND REFORM COMMISSION OF THE PEOPLE'S REPUBLIC OF CHINA

Signed on 16 July 2018 as: "The signing of an agreement on Circular Economy (CE) by the world's two largest economies that could soon accelerate adoption of circular economy practices at a global scale, creating potential for a 'system shift' towards a low carbon, regenerative economy." The next step was for the a new analysis on CE by the Ellen MacArthur Foundation, to be published in December 2019 that finds that a transition to a Circular Economy in China's cities could make goods and services more affordable for citizens, and reduce the impacts normally associated with middle class lifestyles, such as traffic congestion and air pollution.

## 4. Ellen MacArthur Foundation report

The Report was released in December 2018 and done by Arup Corporation (as noted below from its Cover) who are located in the EU and a long-time involved science, technology, engineering and economics company for many regions, nations and cities around the world [4, 5] (**Figure 3**).

One of the critical tasks of the Circular Economy package is the development of innovative product requirements under the Eco-design directive, such as durability and recyclability. In this respect, it is very likely that the Commission will adopt a proposal to differentiate fees by producers in Extended Producer Responsibility schemes according to the real end-of-life costs and recyclability of their products.

As part of the regular reviews of BAT (Best Available Techniques), the circular economy package should also include guidance on best waste management and

Arup Corporation Report on Circular Economics for Ellen MacArthur Foundation

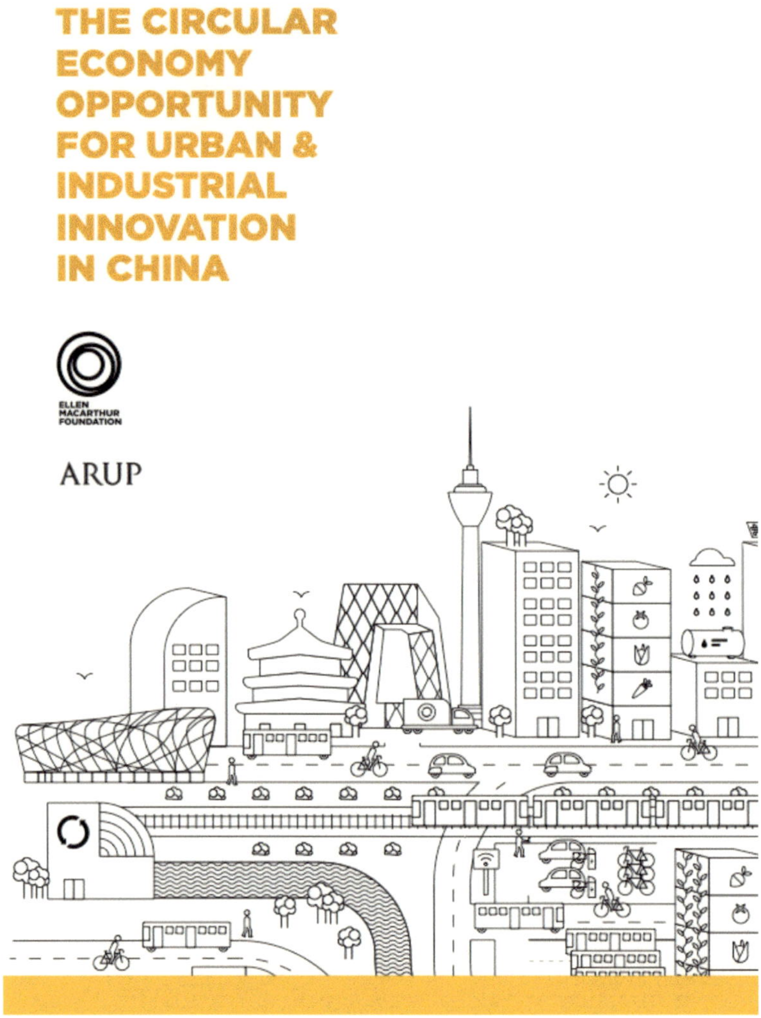


**Figure 3.**
*Copyright: Ellen MacArthur Foundation.*

resource efficiency practices for production processes in industrial sectors, improving the uptake of the European Eco-Management and Audit Scheme (EMAS) and the environmental technology verification system as well as methods to evaluate and make decisions on products (health, environment and nature ingredients and more) such as "earth accounting" has started to do within the circular economy paradigm (http://wwwearthaccounting.com).

What is critical now is that China sees the CE as part of its move into The Next Economics [6] due to Qualitative Economics (2008 and now 2019) playing a significant role in how economics works and should work (**Figure 4**).

The reference models underlying the circular economy stem from the collaboration between the industrial design strategist, William McDonough (Cradle to Cradle (C2C)) and the German chemist Michael Braungart, both collaborators of the Ellen MacArthur Foundation.

One of the key priorities for Europe in the circular economy is that of resource efficiency and waste reduction (Zero Waste Strategy). In this field several *stakeholders* are active, such as Remedia, the collective system of manufacturers of electrical

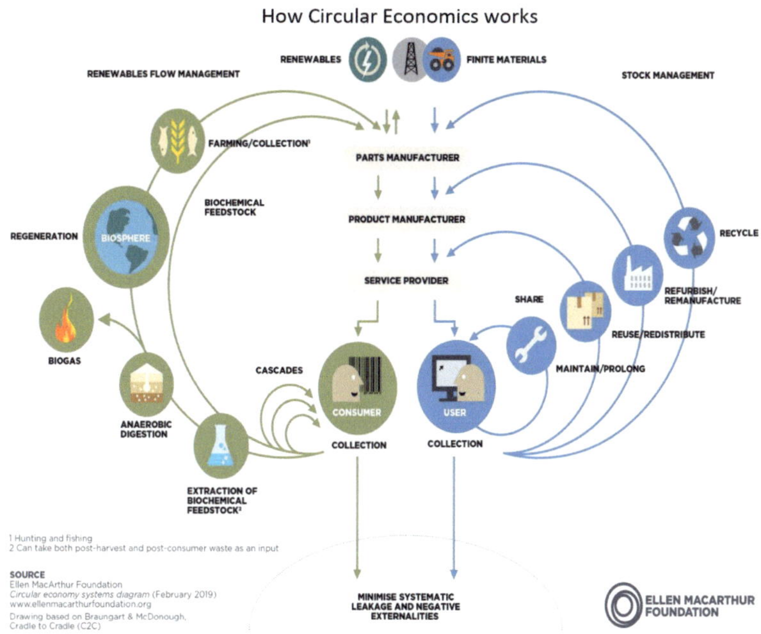


**Figure 4.**
*Ellen MacArthur Foundation plan.*

and electronic equipment that is developing a long-term vision to fully exploit the potential of secondary raw materials and to increase the innovation capacity of the recycling sector in the EU [7].

There is also an absolute need to improve the awareness of European industry in terms of access to critical raw materials, thanks to a closer monitoring of the global markets of non-energy resources. Too many companies ignore the fact that their products are threatened by the risk of future, reliable and fair supply, of strategic resources.

## 5. Economics must be both Qualitative and Quantitative

When economics is referred to as a science, it must act and perform as science which is both qualitative and quantitative. There are two critical factors to follow:

a. *Economic importance*: This analysis is obtained by evaluating the share of each material associated with macro industrial sectors at European level. These quotas are therefore related to the gross value added of the macro sectors compared to the GDP of the EU. The value obtained is then evaluated on the basis of the total EU GDP in order to define an index of global economic importance for a given material.

b. *Supply risk*: The World governance indicator (Wgi) is used to measure this risk. This indicator takes into account a very wide variety of criteria, such as levels of responsibility, political stability, absence of violence, government effectiveness, quality of legislation, rule of law and control of corruption.

Minerals and critical metals are essential for environmental technologies such as solar photovoltaics, wind power, lighting and for low-carbon electric vehicle industries. Interruptions in the supply chain in times of crisis can be dramatic from the environmental, social and economic point of view. This is the main reason why there is a need for a global framework in terms of access to raw materials, capable of overcoming the traditional neoclassical model of supply and demand.

While the science of chemistry is important, the issue is to be able to recycle and reuse the plastic that is made and used for bottles, packaging and more. Now, the plastic can be gathered from reusable products that provide a resource from which to make new products (**Figure 5**).

In a circular economy, buildings would be modular, durable, and flexible. The benefits of digitizing the built environment would go beyond improving energy efficiency and enhance productivity overall. Embracing a circular built environment would reshape both asset utilization and material management in the sector. Inhabitants would enjoy better indoor and outdoor air quality.

The qualitative leap can only be made if we know how to acquire in-depth knowledge about the availability of secondary raw materials. It is also necessary that the actors of the system (raw materials industry, end users, institutions, companies and consumers) collaborate closely to achieve common reuse and recycling targets along the entire production and distribution chain. In this way consumers will know exactly what they are buying and the impacts and potential for re-use and recycling of the products they have chosen.

Moreover, collection systems are still too expensive and inefficient which does not help industrial companies to abandon the traditional production systems based on the linear (flat economic) transformation of materials into products and their disposal once they are consumed. Therefore, the Commission is considering the possibility of introducing further simplifications to promote increased efficiency of collection systems through the circular economy paradigm. Hence by integrating these systems with the upstream industries that make use of recycled components and raw materials from products entering, the end of their life stage can be profitably met as well as protecting the environment (**Figure 6**).

Case of Plastics from the MacArthur Foundation

**Figure 5.**
*Copyright: Ellen MacArthur Foundation.*

*DOI: http://dx.doi.org/10.5772/intechopen.88617*

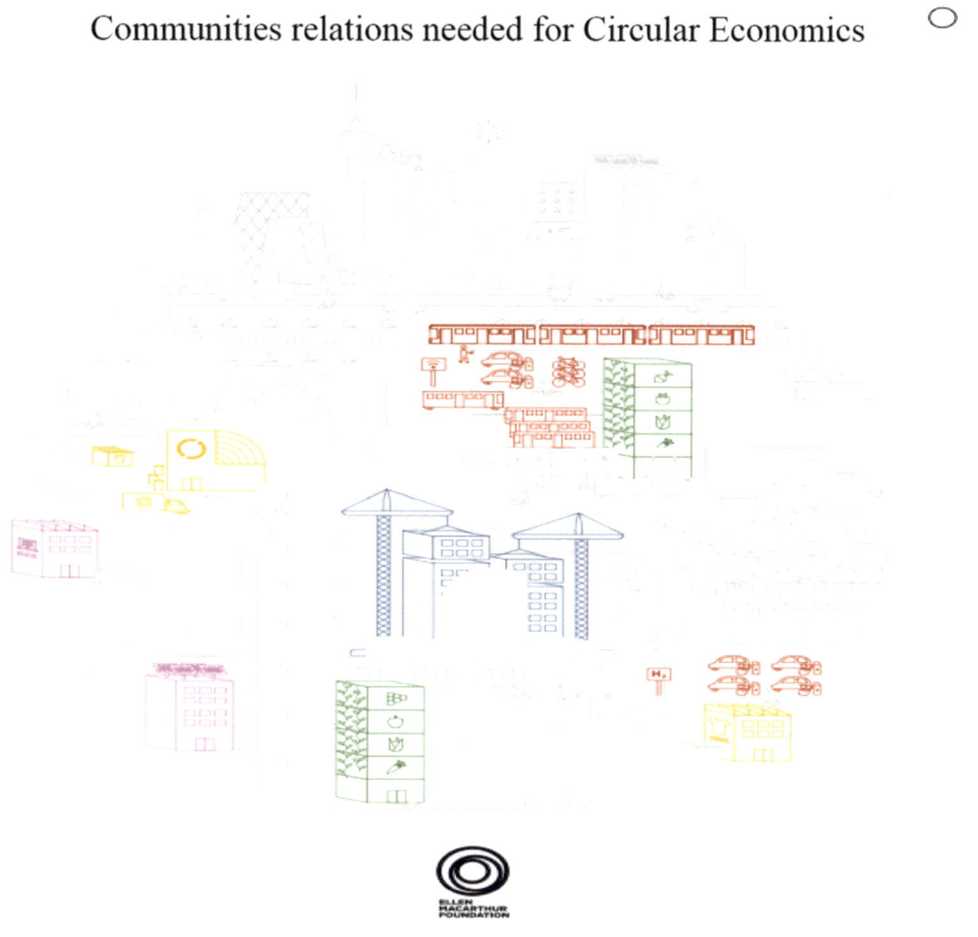


**Figure 6.**
*Copyright: Ellen MacArthur Foundation.*

Many modern products, especially those of high-tech media and those that transit the web and social networks are truly global, from their creation to distribution and sales. Tools like Netflix, YouTube, Facebook and other forms of communication must be part of a circular economy model that directs towards the reuse and recycling of products in order to create new ones. The whole process can and must be improved.

## 6. Conclusion as next steps

Success will only be seen when we are able to achieve a series of improvements over access to and availability of raw materials, with sustainable resource management within a greener, more circular economy and global eco-innovations in technology both in commercial practices to meet the growing demand and ensure the needs in terms of raw material supply. This effort should also lead to changes in behavior that are more consistent with consumers' sustainable use of raw materials.

The dynamics linked to the circular economy can thus be transformed into a solid pillar of the growth strategy of the European Union. The world of industry, in collaboration with governments, must work to address social environmental challenges and generate concrete benefits for individuals and their communities, especially for families and relatives today which is tomorrow now.

Above all, Circular Economics is global, and is NOT based on supply and demand. Nations, regions, cities and communities all need to be joined together as they: (1) work to reverse climate change, (2) create new business models and systems, (3) develop new areas of work, business and family, and (4) create areas that are needed now and into the future for ALL people around the world.

## Author details

Woodrow W. Clark II[1*] and Danilo Bonato[2]

1 Founder and Managing Director, Clark Strategic Partners, Beverly Hills, CA, USA

2 Director General Remedia, Milan, Italy

*Address all correspondence to: wwclark13@gmail.com

## References

[1] Clark WW II, Bonato D. Chapter 7: Circular economics. In: Clark WW II, Fast M, editors. Qualitative and Quantitative Economics (Q2E): Making Economics into a Science. New York, USA: Hauppauge, Nova Press; 2019. Available from: www.novapublishers.com

[2] Clark WW II, Bonato D. Case: Circular economics. In: Clark WW II, Fast M, editors. Qualitative Economics. 2nd ed. Cham, Switzerland: Springer Press; 2019

[3] Clark WW, Bonato D. Circular Economics. HuffPost; 2015. Available from: https://www.huffingtonpost.com/entry/circular-economy_2_b_7029130

[4] Arup Corporation Report. The Future of Urban Consumption in a l.5 C World: C40. 2018

[5] Ellen MacArthur Foundation, Gravis L, editor. @EMacArthurFR. Cowes, Isle of Wight, UK; December 2019. Available from: ellenmacarthurfoundation.org/circulareconomy.com

[6] Clark WW II, editor. The Next Economics: Global Cases in Energy, Environment, and Climate Change. Springer Press; 2013

[7] Bonato D. General Manager. Remedia. Milan, Italy. Available from: danilo.bonato@consorzioremedia.it

Chapter 9

# Green Technology Innovations Development in China: Trend and Application

*Xiaodong Lai*

## Abstract

This chapter aims to explore what subjects have been addressed in green technology innovation (GTI) in China and initiate a journey for the next generation of sustainable-oriented research. Thus, the work examined the literatures enlisted in the database of China National Knowledge Infrastructure (CNKI) on the topics of GTI from 1994 to 2019. Some critical discussion and conclusion are sighted as follows: (1) the research of GTI in China is getting mature compared to the energy-innovation related topics and the researches 10 years ago. It is becoming a dominated research subject. (2) The qualitative publications dominates the researches, the empirical researches are in a shortage. (3) The research subjects are multi-perspective and multi-disciplinary, covering environment science, management, energy and fuels, economics and social behavior. New vibrancy of advanced theoretical and methodological research is particularly needed. (4) The trend of green technology research appears an interdisciplinary research with the themes related to environmental subject, science technology, business economics, engineering and energy & fuels. (5) Different policy implementations have different effects due to the cost structure and maturity of renewable energy. (6) GTI cannot be isolated from the policy or regulation regime, and is becoming a new underpin of current sustainable development in China.

**Keywords:** green technology innovation, technology innovation, management, sustainable development

## 1. Introduction

The global warming, frequent natural disasters and resource shortage occurred in the twenty-one centuries are forcing people to excogitate a new way to save our earth. Many countries are focusing on the development of low carbon economy or green economy. The development with green technology innovation orientation and policy regulation to drive an energy evolution and the establishment of a new economy development with less GHG emission are acknowledged to prolong the climate change [1].

In order to understand the past, present and future with regard to the technology innovation in green research and practice in China, a typical emerging country in Asia, this chapter took the source of primary database of China National Knowledge Infrastructure (CNKI)[1] as a particular intellectual domain for analysis. The purpose of this article tries to provide with facts that help researchers and practitioners understand what issues or subjects have been addressed in green technology innovation and anchor the trends for the next generation of sustainable-oriented research.

The remaining parts of this article are divided into four sections. In section two, the methodology and search selection are presented. In the third section, the analysis of the literatures by classifying them into different types is conducted. In the fourth section, an extensive detail study on the international research themes of green technology innovation is presented, and to a further step. A brief discussion with conclusion is given in the last section.

## 2. Methodology and search criterion

Through our exploration of the literature reviews within the domain of social energy system and sustainable development, we found that the literature review from the perspective of low carbon-oriented green technology is very few [2]. Carbon emission problem is becoming a serious issue right now, which is threatening the welfare of human beings. The research on carbon emission problems is being the mainstream in the existing green technology research. Therefore, this paper continue the research exploration as paper [2] but focus on the development trend in China to further invest the research status of green technology innovation after 10 years.

This article employs a methodology to reviewing the articles cited in the databases "CNKI" with "green technology innovation" as the "topic." The earliest published article related to green technology innovation topic appeared in 1994 [3]. By pulling all the articles from 1994 to 2019, 2014 articles are identified that fell within the domain of our topic "green technology innovation". Eleven overlapped article has been omitted, we has an overall glanced over for all the retrieved publications, and removed the reports, notices, announcements, conference summaries, exclusive interview, laws or regulation introductions and some researches with no green technology innovation involved, we also deleted some anonymous researches or some companies' green technology introductions. In addition, by scanning the titles and the abstracts of each article published from 1994 to 2019 and using related keywords for double review, it was found that 1348 articles mainly focus on our themes of "green technology innovation", thus, 1348 articles are kept as the research sample. Considering the searching engine objective problem, some articles with such subject may not 100% be retrieved, Therefore, this paper modestly believed that this approach was likely to have presented nearly every related article in these two databases.

It should be noted that, we cannot possibly provide a truly comprehensive review for all the articles, especially for those in a particular research field, e.g. chemistry and

---

[1] CNKI: CNKI project started with an e-journal product and later further expand the product line to cover newspapers, dissertations, proceedings, yearbooks and reference works and etc. CNKI is a symbol of Chinese e-publishing industry, which greatly boosted the Chinese library systems to go digital and helped researchers with their work. So far CNKI academic databases have been serving more than 5500 universities, public and corporate libraries, hospitals and other institutions inside and outside of China.

eco-biology field. However, we do feel that many of the interesting insights arise through a detailed review of these articles. In our research we organized the articles with six different approaches by (1) period sequence, (2) research methods, (3) research level, (4) research subjects, (5) keywords cluster, (6) Institution, (7) Foundation, and (8) themes of articles (this is separately analyzed in the fourth section).

With the above approaches used in this article, this paper is trying to propose an indication of the trend of green technology research and help the readers to understand the milestone throughout the development of green technology innovation. It can also serve the on-site practitioners to understand the research trend in China.

## 3. Overview of the literatures from the database of CNKI

### 3.1 Publications classified by period sequence

Firstly, the raw data retrieved from the database of CNKI with timelines and publication quantity shown in **Table 1** and made a time series bar graph in **Figure 1**. It shows the yearly publication distribution in the field of green technology innovation each year. Obviously, in the period from 1994 to 2000, there were a few articles related to green technology innovation published internationally. Only 50 articles appeared per our review. However, in the period from 2001 to 2005, it was 176 articles, three times more than the previous. Late on, in the period of 2006–2010, a dramatic increase of information consisting of 251 articles, it remains a steady publication increase as well as green concept are recognized by the public, and the QTY increased by 922 publications in the period of 2011–2019 as of 7-31-2019.

It can be seen from the above trends that the academic field is paying more and more attention to the innovation of green technologies. It indicates that the new research areas of green technology innovation mainly focus on technology development and model innovation. The new trend that focuses on technology development represents the new requirements of the modern society for the innovation level, that is, the hard demand for new technology development to improve the relevant industry benefits. The combination of academic research and social production is more closely integrated.

### 3.2 Publications classified by research method

A further review of these articles from the research method perspective was conducted. Referring to the literature of Shi and Lai [2], we divided the research methods into four categories: conceptual, model, empirical and qualitative methods. Here we only explore the articles cited in databases "CNKI" and selected the high cited articles and hot articles from them. In this way we got 1346 articles for analyzing. **Table 2** and **Figure 2** shows the distributions of those articles across conceptual, model, empirical and qualitative methods in every 5 years and the total quantity from 1994 to 2019 per each method. It can be seen that the model articles have gained 18% and the empirical articles have gained 15%. While the conceptual account for 49%, and the qualitative method occupies 18%, both achieve 67%. It indicates that the green technology innovation in China is still at a developing stage and the method of model and empirical are not adequate. It needs to encourage researchers to use data and models for better illustrating the relationship between elements in order to adapt to the needs of green technology development.

| Year | 1994 | 1995 | 1996 | 1997 | 1998 | 1999 | 2000 | 2001 | 2002 | 2003 | 2004 | 2005 | 2006 | 2007 | 2008 | 2009 | 2010 | 2011 | 2012 | 2013 | 2014 | 2015 | 2016 | 2017 | 2018 | 2019 |
|---|---|---|---|---|---|---|---|---|---|---|---|---|---|---|---|---|---|---|---|---|---|---|---|---|---|---|
| QTY | 2 | 1 | 3 | 5 | 12 | 13 | 14 | 26 | 32 | 32 | 37 | 49 | 26 | 61 | 53 | 38 | 73 | 45 | 64 | 71 | 82 | 96 | 164 | 150 | 172 | 78 |

*The cut-off date for the above data is 7-31-2019.*

**Table 1.**
*Literatures publication status on green technology innovation from 1994 to 2019.*

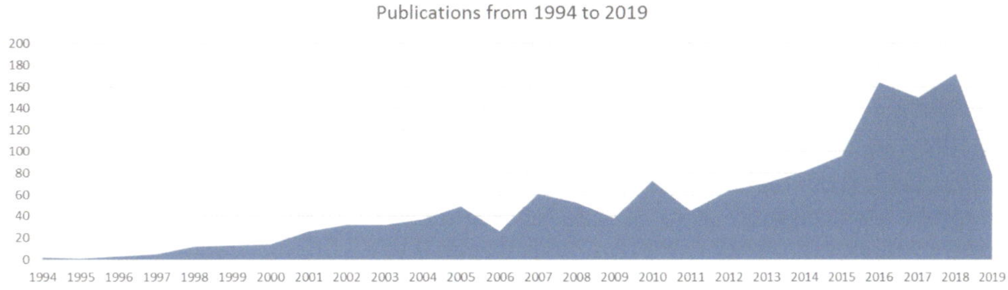


**Figure 1.**
*Literatures on green technology innovation publication status from 1994 to 2019.*

| Types | 1994–1998 | 1999–2003 | 2004–2008 | 2009–2013 | 2014–2019 | Total | Rate |
|---|---|---|---|---|---|---|---|
| Empirical | 1 | 2 | 10 | 35 | 154 | 202 | 15% |
| Qualitative | 6 | 17 | 40 | 75 | 109 | 247 | 18% |
| Conceptual | 26 | 86 | 188 | 149 | 210 | 659 | 49% |
| Model | 1 | 6 | 8 | 35 | 188 | 238 | 18% |
| Sub-total | 34 | 111 | 246 | 294 | 661 | 1346 | |

**Table 2.**
*Percentage distribution of literatures by research methods in every 5 years.*

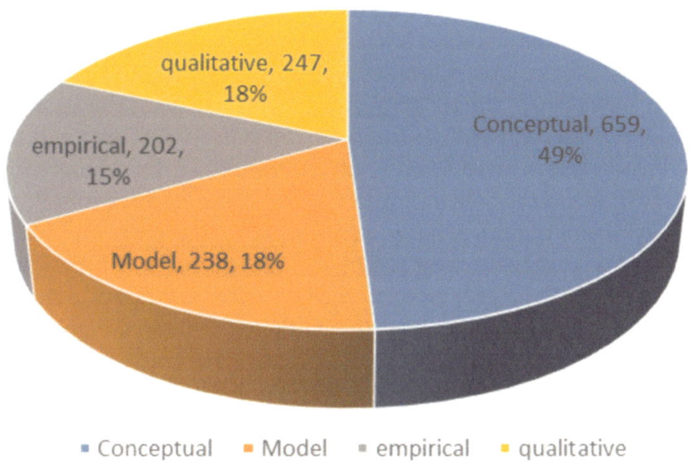


**Figure 2.**
*Percentage distribution of literatures by research methods.*

In fact, the model method can help apply the technology innovation influence on practice. The most representative publication using formal model method is Zhang and Zhu [4], the paper chooses productivity of resources and environment loads as output variables, and develops an empirical study using four-stage DEA model to analyze technological innovation efficiency of industrial enterprises, result suggest that the environmental factors are conductive to the technological innovation efficiency. It has been cited more than 162 times as of July 31, 2019. The other four publications with higher citation are Xu et al. [5] (109 times citation), Zhang and Qiu [6] (85 times citation), Qian et al. [7] (74 times citation) and Luo and Liang [8] (68 times citation).

The empirical research method is a powerful way to analyze the relation of different factor. For instance, Li. et al. [9] uses none radial and none angle SBM efficiency measurement model considering the undesirable outputs, combined with the ML productivity index to measure the efficiency of green technology and green total factor productivity of industry, which gains 283 times citation as of July 31, 2019, which is the highest citation publication so far. Another three authors [10] use empirical research method to analyze the relationship among market demand, green product innovation, and firm performance. In addition to technology adoption and field experiments in different industries, the empirical method is used to research the impact of low carbon technology policies and the application of low carbon technology in the market. There are four typical representative publications with higher citation (over 60-time citation), namely, Chen [11] (111 times citation). Chen et al. [12] (86 times citation), Li, et al. [13] (76 times citation) and Wang and Zhu [14] (64 times citation).

The qualitative research of green technology innovation focuses on the technology introduction, adoption and green path discussion. For example, Guo [15], a literature with more than 125 times citation as of the end of July, it believed that ecological industrial park is a concrete path to the realization of the sustainable development of industry economy and the most ideal model to realize the sustainable development of whole society. Dai and Liu [16] thought that green innovation in China sustainable development need the driving forces from demand, institution and technology innovation, an environmental innovation system from national perspective is needed. It gains 93 times citation as of July 31, 2019, the other representative articles are Wu and Yang [17], Chen [18] and Hua [19].

The research employed a conceptual method to introduce green technology concept from different perspectives such as innovation concept, new technology application, policy and etc. For example, the representative authors Qin and Yang [20] introduced Xi JinPing's theory of green development on the basis of the worldwide trend of green growth. It includes the following aspects: transforming the economic development mode poses the premise to realize the green growth; developing recycling economy is an important means of promoting it; improving green technology offers technical support for it; handling with the relation between developing economy and protecting ecological environment is a basic requirement for promoting it; advocating green consumption is the important way to promote it; improving the living environment for the people is the ultimate goal for China to choose it. Xi's thought of green growth is of great theoretical and realistic significance for China green development. As of July 31, 2019, this article gains more than 106-time citation. Zhong and Wang [21] thought that green technology innovation is an effective method to resolve the contradictions between enterprise economic development and environmental deterioration, and proposed related recommendation on the establishment of green technology innovation system, this article has been cited with more than 97 times as of 2019. Another two outstanding representative articles are Xu and Wang [22], Zhao [23], these articles introduce the latest development in green technology innovation.

From above analysis, we can see the green technology innovation in China is still under development stage, the conceptual and qualitative methods researches dominate in quantity for quite some time, while the model and empirical researches need more practice and employment.

### 3.3 Publications classified by research level

We reviewed all the articles' main contents based on the abstract descriptions. For the analysis purpose, we took the research level code scheme based on

Schumpeter's micro-meso-macroeconomics definition [24], and classified the research level as "macroscopic, mesoscopic and microscopical". For the articles regarding policies, regulations, national mechanism, national ecosystem, global comparison, concepts and sustainable development etc., we coded them as a macroscopic level. While for the articles involved with the country or territory's economy development, the country or territory's technology adoption and diffusion, national policy or technology promotion, regional economic development or technology adoption, and etc., we coded them as the mesoscopic level. For the articles involved with concept, basic research, product design, risk management, empirical research, models, entrepreneurship innovation, and technology innovation capability, we coded them as the microscopical level.

**Figure 3** shows the distribution of the different research levels, macroscopic research level gains 54%, the research related to national policies, including the impact of different policies on different technological innovations, the assistance of policies to renewable energy technologies entering the market and so on, are all gained much attention by the Chinese scholars. The top five cited articles of this level are Guo [11], Zhao [25], Chen [18], Zhou [26], Li and Yang [27] mesoscopic research level are keeps 11%, the themes related to the impact of green revolution on industry and territory, including the acceptance of green economy and new technologies in society, the efficiency of new technologies in some industries and so on. For example, Lin et al. [28] analyzed the green technology innovation efficiency of China's manufacturing industries. Wang et al. [29] discussed the green development strategy in Peal River Delta of China, and Luo and Liang [8] studied the regional industrial enterprises green technology innovation efficiency and factors decode. You and Wang [30] verified the environmental regulation effectiveness on R&D bias of strengthening the green technology.

Microscopic research level are keeps 35%, this phenomenon reflects that current researches still focused on specific and concrete aspects, especially the technology innovation and most of them are about the application of new technology in the firms and environmental improvement. Most of the articles in microscopical level are about the inventions of new technology and method, including the technique of detecting chemical substances, improvements in technologies and so on. Typical representative articles are Wang et al. [31], Wang and Li [32], Zhang and Li [33], etc.

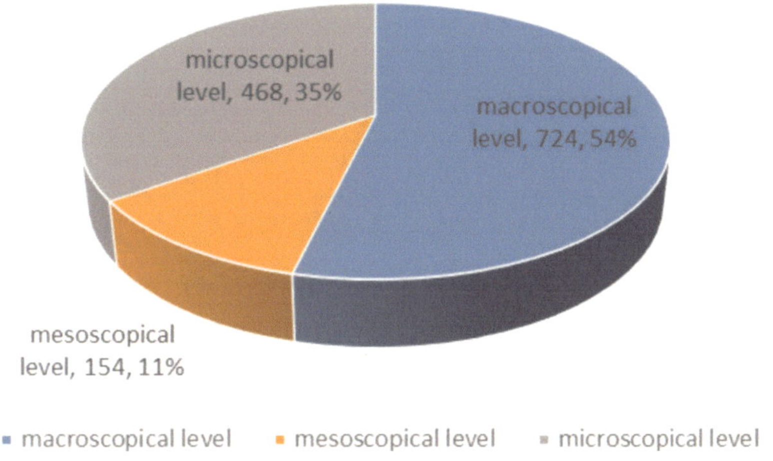


**Figure 3.**
*Literatures on green technology classified by research level.*

### 3.4 Publications classified by research subject area

We classified these 1346 articles per research subjects. Based on the original analysis chart downloaded from the CNKI data base, it has total 32 research subjects involved with green technologies. Some of the articles are multidisciplinary, for example, Cao and Zhang [34], Li [35], Zhang et al. [36], Zhang et al. [33]. In order to have a better understanding about the actual classification, we consolidated the overlapped research subjects. For the subject with one article, we put it into the "others" portion for the subject with less than 10 publications, we simplified and showed them in **Table 3**.

From **Table 3**, it can be seen that the distribution of research subject are: green technology innovation (301 articles, 21.41%), enterprise management (183 articles, 13.02%), green technology (111 articles, 7.89%), technology innovation (75 articles, 5.33%), sustainable development (72 articles, 5.12%), green innovation (70 articles, 4.98%), China (64 articles, 4.55%), sustainable/green consumption (58 articles, 4.13%), environmental regulation (55 articles, 3.91%), green development (51 articles, 3.63%). The rest of the articles' distribution is shown in **Table 3**.

The trend of green technology research appears an interdisciplinary research with the topics related to environmental subject, science technology, business economics, engineering and energy & fuels, which are accounting for 83% of the total. It is worth mentioning that the research articles of the top two areas–environmental sciences ecology and business economics are much more compared with other industries, the top five areas in the review of paper [2] is about for 67% of the total. In a word, the trend in the interdisciplinary field remains unchanged but with the increase in social participation, it is relatively concentrated in several subject areas (**Figure 4**).

| Subject | QTY | Rate of 1346 (%) | Subject | Qty | Rate of 1346 |
|---|---|---|---|---|---|
| Green technology innovation | 301 | 21.41 | Ecological civilization | 39 | 2.77 |
| Enterprise Management | 183 | 13.02 | Green building | 25 | 1.78 |
| Green technology | 111 | 7.89 | Green industry system engineering | 22 | 1.56 |
| Technology innovation | 75 | 5.33 | Greenization | 22 | 1.56 |
| Sustainable development | 72 | 5.12 | Green construction | 20 | 1.42 |
| Green innovation | 70 | 4.98 | Environmentally conscious products | 19 | 1.35 |
| China | 64 | 4.55 | Green barrier | 18 | 1.28 |
| Sustainable/green consumption | 58 | 4.13 | Fiscal management | 18 | 1.28 |
| Environmental regulation | 55 | 3.91 | Green transformation | 17 | 1.21 |
| Green development | 51 | 3.63 | Influence factor | 15 | 1.07 |
| Ecological or green economy | 51 | 3.63 | Green marketing | 15 | 1.07 |
| Circular economy | 46 | 3.27 | Others | 39 | 2.77 |

**Table 3.**
*Literatures on green technology classified by research subject.*

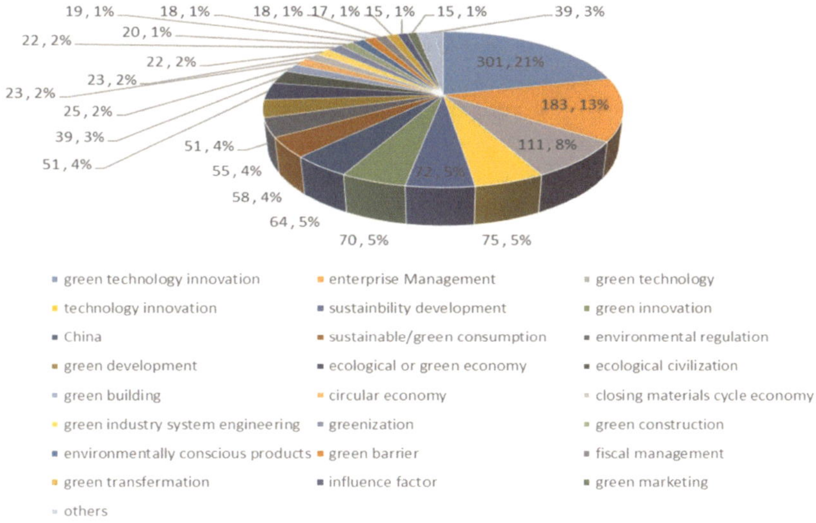


**Figure 4.**
*Literatures on green technology classified by research subject.*

## 3.5 Publications classified by keywords cluster

Based on the above research subject area, we seek further at the research trajectory of scholars from keywords cluster. Firstly, we searched core journals with the theme of "green technology innovation" in CNKI database and found 2014 relevant literatures[2]. Then, we imported the data into Citespace for keyword co-occurrence analysis, obtained the atlas, and sorted the keyword frequency into a table. As can be seen from **Figure 5** the keywords with high frequency are "Innovation", "Technology", "Policy", "Sustainability", "System", "Performance", "Management", "Energy", "Climate change", "Model" etc. We decided to further analyze them.

Technological innovation-related keywords gain highest occurrence frequency. In recent years, energy shortage, climate change, environmental degradation, green economy, low carbon and other phenomena have made people more aware of the

**Figure 5.**
*Keywords cluster with occurrence frequency.*

[2] Here, the authors didn't screen the publication as above mentioned 1346 article.

importance of sustainable development, and technological innovation is an important driving force for sustainable development. Chinese government conducted various policies and adjusted industrial structures to promote green innovative development of technologies, management and institutions, improve the efficiency of energy consumption, and support green innovative activities. For example, the renewable energies [37] like new energy [38, 39] biotechnology, solar energy, water and other green energy, the green innovations in process control, workmanship [40, 41] or construction method [42], those are the key areas for our researcher to explore and develop.

### 3.6 Publications classified by institution

We categorized the 1346 publications according the institution/organizations with a time series. Related results are shown in **Figure 6**, it can be seen that the top institution is Zhejiang University who gains 36 articles, the second one is HUST (Huazhong Science and Technology) with 35 articles, then Northeastern University, Harbin Engineering University of Science and Technology and Kunming University of Science and Technology. There are 30 institutions total with above 10 articles contribution, and most of the institutions are science and technological universities or comprehensive universities with technological and environmental subjects. Those institutions with high publication are supported by the National Nature Science Found and Social Science Found.

### 3.7 Researches in international journals by Chinese scholars

It is known that the setup of green technology innovation is from the western's countries like European and American counties, who dominated the technologies and energy consumption at the beginning, then turns to share market of developed countries and developing countries. Share of developing countries escalates and develops, China is one of the most powerful countries among them. However, with the high-speed development of economic and social development, technology innovation changing rapidly and increasing living standards of people, the consumption and waste of energy resources are continuously compounded as well. As the biggest developing country, China has the biggest environmental protection market in the world, thus, this paper also screened the international publications written by Chinese Scholar. **Table 4** presents 10 documents originated from Chinese scholars by browsing the theme of green technology in web of science, an international database for worldwide researchers. It can be seen that green innovation research themes of Chinese scholars mainly concentrate on fields related to

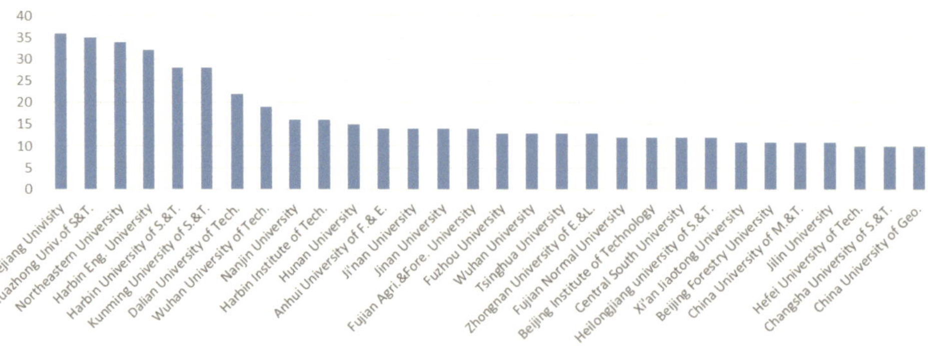


**Figure 6.**
*Green technology publications per institution.*

| Code | Headline | Author | Subject theme | Method |
|---|---|---|---|---|
| 1 | Sustainable energy development and climate change in China [43] | Ren, Zeng, Zhou | Environment and ecology | Qualitative analysis |
| 2 | Integrated management for renewable energy resources and CDM resources: A case study [44] | Jing, Mingshan | Business and economy | Case method |
| 4 | Scenario analysis on alternative fuel/vehicle for China's future road transport: Life-cycle energy demand and GHG emissions [45] | Ou, Zhang, Chang | Energy and fuel | Modeling method |
| 5 | Study on China's low carbon development in an Economy-Energy-Electricity-Environment framework [46] | Hu, Yuan, Hu | Energy and fuel | Qualitative analysis |
| 6 | Network Environ Perspective for Urban Metabolism and Carbon Emissions: A Case Study of Vienna, Austria [47] | Chen, Chen | Engineering | Modeling method |
| 7 | Influence of household biogas digester use on household energy consumption in a semi-arid rural region of northwest China [48] | Ding, Niu, Chen, Du, Wu | Energy and fuel | Investigation method |
| 8 | An optimization model for renewable energy generation and its application in China: A perspective of maximum utilization [47] | Cong | Energy and fuel | Literature method |
| 9 | A review of clean energy innovation and technology transfer in China [49] | Liu, Liang | Energy and fuel | Literature method |
| 10 | A review of China's approaches toward a sustainable energy future: the period since 1990 [50] | Zhu, Zhuang, Xiong | Energy and fuel | Literature method |

**Table 4.**
*Document researches in selected national journals by Chinese scholars.*

energy, fuel, commercial economy, environmental sciences, engineering and so on, while researches on the theory exploration and management level of energy development have not embodied yet. The Chinese scholars' publications in international journals are less than in the developed countries or regions, but more than the publications from other developing countries.

## 4. Publications classified by major themes

Our main subject discussed here is for the articles related to green technology innovation, which is more specific and concise restrict in green field than the normal innovation. However, this subject also has the similar area applied to product innovation, technology innovation, technology transfer, technology diffusion, regulation or policy innovations, innovation abilities, even the innovation to the individual firms or organizations. This is a little different from the subject category Based on the characters of green technology innovation, we decompose the broader subject and code the research themes of green technology into seven themes. This is taken the reference of the article by Shane and Ulrich [51] with some modification: (1) regulation or policy innovation, (2) technology innovation adoption & diffusion, (3) technology transfer, (4) technology innovation capability, (5) basic research and advance development, and (6) Entrepreneurship innovation (see **Table 5**).

We went over all these 1346 articles' abstract contents and categorized the articles based on the above code scheme and screened some similar publications.

| Major themes | Sub-themes |
|---|---|
| 1. Regulation or policy innovation | 1.1. Effect of innovation on economic growth |
| | 1.2. Factors influencing the rate of innovation |
| | 1.3. Tools used by policy maker |
| | 1.4. Impact of specific policies |
| 2. Technology innovation adoption and diffusion | 2.1. Pure technology Introduction |
| | 2.2. Technology adoption method introduction |
| | 2.3. Technology diffusion introduction |
| 3. Technology transfer | 3.1. Patent |
| | 3.2. Learning |
| | 3.3. Technology spillover and policy impact |
| 4. Technology innovation capability | 4.1. Management innovation |
| | 4.2. Design innovation |
| | 4.3. Process innovation |
| | 4.4. Organization Innovation |
| | 4.5. Innovation assessment |
| 5. Basic research and advance development | 5.1. Conceptual/theory |
| | 5.2. Review |
| | 5.3. Framework/models |
| | 5.4. Driving mechanism |
| | 5.5. Risk innovation and management |
| 6. Entrepreneurship innovation | 6.1. Green design strategy |
| | 6.2. Innovation efficiency |
| | 6.3. Innovation behavior |

**Table 5.**
*Major themes and sub-themes code scheme with the domain of green technology innovation.*

69% articles are discussed the theme of regulation or policy innovation, 14% articles were involved with the theme of policies. 3% percent of the articles are discussed with the theme of technology adoption and diffusion. The articles contented with the theme of technology transfer gained 6%. While the articles conducted with the theme of technology innovation capability occupied 6% either. The rest 2% fell on the theme of entrepreneurship Innovation. Detail distribution is shown in **Figure** 7. A further detail analysis by different themes category was illustrated in the Section from 4.1–4.6. For research purposes, we also extended some authors literatures involved in this type of theme from the deeper perspective and across research perspective instead of more than 1346 articles only we stated.

## 4.1 Regulation effect on green technology innovation

The theme of regulation or policy is one of the main and traditional research areas of technology innovation. It has a critical impact on innovation theory development. We classified the articles which related to policy into four categories the effects of innovation on economic growth, factors influencing the rate of innovation, policy tools used by policymaker and the impact of specific policies.

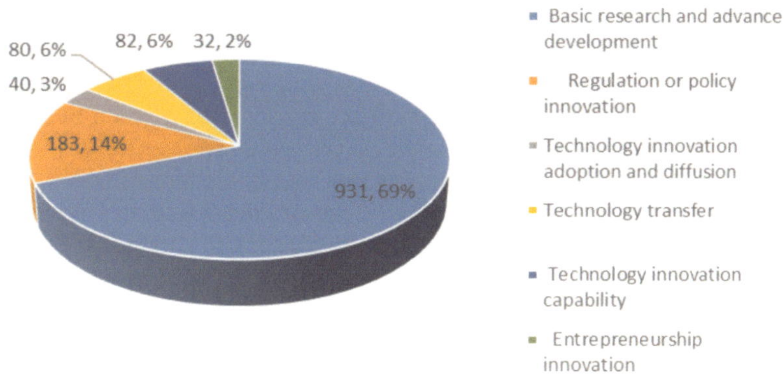


**Figure 7.**
*Publication on green technology innovation classified by themes.*

It has 48 articles involved with the effect of innovation on economic growth, there are some typical articles like Hou and Su [52], who firstly studied the green barriers effect to the exported industrial technological innovation. Peng and Sun [53] explored the challenges and strategic solution for the green economic development in China. Li et al. [9] reviewed the environmental regulation effect to the total factors of green production efficiency improvement. Feng et al. [54] discussed the relationships among the regulation difference, innovation driven and economy development in China and put forward the related recommendation regarding the effect of regulation. Xu and Zhen [55], Pei et al. [56], both publications explored the effectiveness of regulation on the economy development in Long River Belt in China. Wu and Yu [57] had a further study regarding the environmental regulation impact on the production efficiency improvement and technology innovation.

71 articles studied the factors influencing the rate of innovation. Li [58] and Gao and Wang [59] conducted the green innovation efficiency of high energy consumption industries in Jing-jin-Ji district from a special perspective for China industries development. Yang et al. [60] discussed the green innovation impacts in China and pointed that the regulation factor is one of the critical items for green development. While the rate of innovation is multiple [61, 62].

Six articles discussed the tools used by policy maker, the representative authors are Li et al. [63], Wang et al. [64], Wei [65] and Shi [66].

The theme of impact of specific policies has 59 articles, most of the publications are concentrated on the environmental regulation effect on the green innovation performance or green economic development. The most outstanding articles are Zhang et al. [67]; Li et al. [68]; Xu and Wang [69]; Francesco et al. [70]; Zhang and Qu [6]; Xu et al. [5] and Wang [71].

Throughout the above articles review, it can be seen that the effects of different policy implementations have different effects due to the different cost structure and maturity of renewable energy. The innovation of green technology can be induced by policies. The feed-in tariffs are relatively more practical. By controlling the relative price of the alternative factors, the demand factors, general scientific ability, the after controlling the relative differences in the economies, the patents tendencies, the number of patents was used as a measure of innovation ability.

## 4.2 Technology innovation adoptions and diffusion

The theme of technology innovation adoption and diffusion is the major impressive body research of green technology innovation for our researchers. A total of 93 articles explored this theme. We divided the theme into three sub-themes: pure technology introduction, technology method or theory introduction and technology diffusion introduction.

Several articles introduced the pure green technology adoptions in different fields such as Hong Kong-Zhuhai-Macao bridge construction, green building, oil and mine industry etc. [72–79]. These articles are good representative example for green technology adoptions in China.

Technology innovation and diffusion theories are discussed and extended. Yun and Lee examined key factors of renewable energy systems diffusion from a socio-technological perspective [80].

The technology adoption theory with empirical application based is also one of the key research themes. E.g., Zhou et al. [81] investigated a specific example of a demonstration project in China to support the large-scale diffusion of green technology and its pilot implementations and revealed that these demonstrations face a different set of diffusion barriers. Zhu et al. [82] studied Chinese manufacturers GSCM adoption practices to see if this adoption affects their performance or not.

The themes about the technology diffusion policy were recommended with the theoretical and empirical analysis in several articles which cover innovative green procurement [83], mechanisms that can accelerate the technology diffusion [84–87].

The above exploration shows that people take great efforts on the real actual experiment and achievement in the adoption of the green technology, such as the sample survey, case study or fields study with primary data. The typical research articles are influential, for example, Zhu and Sarkis [88]. gained more than 108 times citation up to now. "The models that underlie much of the diffusion literature have their roots in physical diffusion processes". Green technology innovation diffusion takes leading the regional economic development, they save the energy with low $CO_2$ emission or zero-energy emission.

Furthermore, this review provides the information that the researchers' focus is within new energy introduction, renew energy development, new methods, new process improvement and even new conception or culture implantation. This is a new trend of green technology development.

### 4.3 Technology transfer

When introducing the technological track of the green revolution, Liu and Liang [88] explores potential policies and schemes promoting the transfer of CCS technologies to developing countries, then makes an attempt to understand technology transfer including its benefits, barriers, and definition. Ai et al. [89] studied the impact of various technological progress patterns on China's regional environmental performance using spatial econometrics and find that there are significant spatial effects of technology innovation, technology transfer on China's regional environmental performance. Liang and Luo (2019) studied the dynamic effect analysis for international R&D capital output to green technology [90].

The realization of technology transfer in green revolution is influenced by many factors. The process of green technology transfer needs to take full account of the capacity of enterprises and the policy support systems. At the same time, the increase of independent innovation capacity will play a positive role in technology transfer. From the perspective of historical development, in the environment of rapid improvement of independent innovation ability and intensified international competition, the transfer of green technology in developed countries will have a more positive impact on developing countries. Developing countries should also try their best to improve their technology adaptation level and absorptive capacity to accept the technology transfer of the green revolution in a proactive manner.

### 4.4 Technology innovation management and capability

The theme of technology innovation capability is a little broad. For analysis like the one cited above, we divided it into four sub-themes such as the innovation research of management capability, design capability, process control capability and organization innovation capability.

Six articles argued the innovation management concept related to green technology innovation. For example, Wu [91] discusses the relationship between green supply chain integration (GSCI) and green innovation and pointes that, in order to improve green innovation performance, managers should strive to integrate resources and capabilities among their organizations, suppliers and customers. The managers should constantly pay attention to market demand trends and maintain a close technical network between supply chain partners.

The design method innovation is also a critical research theme in technology. Six typical articles presented herein are very useful for our practitioners to study. Song and Yang [92] reviewed the relations between the financial performance and green innovation, and pointed that green capabilities are much related with the aid of finance. Han and Yan [93] explored how to the innovation capability impact the enterprises value-adding. Li et al. [68] analyzed the relationship among the environmental regulation, R&D investment and enterprises green technology ability. The rest of articles are Li et al. [94], Hua [95] and Liu et al. [96].

Regardless of the technology innovation itself, the subject of process innovation is also well-accepted by our researchers, particularly in the big construction induction of green innovation, e.g., Fu [97] proposed that environmental regulations and corporate characteristics are the most widely studied factors affecting the adoption of sustainable process technologies.

Organization innovation is another sustentation on green technology. It has three sub-themes such as the effort of organization structure, communication patterns and decision making. Only a small number of typical articles relate to the impact of organizational innovation on green technologies technology innovations.

The above analysis provides a possible direction for us that the effectiveness of innovation does not only depend on the technology but also depend on the related responsible stakeholder concept of innovation and its capability across with the management, organization and process innovation. Based on the deep research, it can be caught that, the capacity innovation, ecological innovation, leadership, and technology trajectory will all become important factors influencing green technology innovation. It is a result of core team cooperation and there are a lot of potential in this area for researching.

Through the above literatures, it was sighted that the research of green technology innovation in enterprise management gradually turns from the macro aspect to the micro aspect. After analyzing the green macro strategy of enterprises, scholars begin to pay more attention to micro factors such as enterprise organization, system design and innovation ability. Meanwhile, government policies are becoming an important influencing factor of low-carbon economy and green economy, and the direction of future research.

### 4.5 Basic research and advance development

The Basic research and advance development theme involved the topics of concept or theory, reviews, framework/models, driving mechanism, risk and innovation management. 697 articles discussed the basics research with the themes of conceptual or theory introduction, most of the articles discussed the concept of green technology innovation and the reason why need develop green economy in

China. [53, 98–100] but it needs to pay attention the green innovation system platform construction such as consumption mode [101], innovation system [102, 103], policy supporting system [104, 105], law system [106] and green cultural system [107, 108].

43 articles are reviews and comment on the green technology development, literatures review, trend and forecast discussion in China. Typical articles are Guo and Zhang et al. [109], it pointed out that the "environment-economy" development science and technology is a big dilemma in the process of sustainable development in various countries. It needs to clarify the impacts among the environmental regulation and governmental R&D funding and green technology innovation and study the function of promoting or suppressing. Wang [110] thinks that the technological innovation generated by environmental policy is the key for China achieving its long-term green development, the market driven force for environmental-friendly innovation activities is lacked, whereas environmental policy can provide motivation. Li and Yang [111] employed an analytical framework proposed by Astley [112], who developed a systematic, coherent analysis on the level, phase, theme and depth of domestic green technology innovation literature, and summarized the characteristics of research and calculated the trend of green technology innovation research by connecting it with the time dimension. Wang [113] summarized the external factors and internal determinants of enterprise competitive advantage from the perspective of regional innovation milieus and green technology innovation and pointed that green technology innovation is the internal factor of enterprise competitive advantage, it is an important part of enterprise internal environment. But with multi factors driving [114].

172 articles studied the theory framework and models of green technology innovation development. The main topic of the theory framework are within the scope of regulation effect to the green technology innovation [115–118], innovation system, the relationship among green technology innovation, industrial agglomeration and ecological efficiency [119, 120]; Li, Peng and Ouyang [9] thought that environmental regulation has an effect on the transformation of China's industrial development mode through green total factor productivity, but there exists the "threshold effect" of environmental regulation strength.

43 articles reviewed the driving mechanism of green technology, Vicky [121] alternative policy mechanisms based on a simulation for green technological innovation, Yu [122] proposed a combination of alliance for enterprises green innovation based on a business case study. Cao (2008) analyzed the interaction among the structure mode, operation mechanism and technology from a perspective of recycling economics, then pointed that that industrial structure transformation with technological innovation is critical. Zhang and Sun [123] conducted an economic-social-recycle system to explore the innovation efficiency mechanism based on an example in Han River eco-economic belt in Hubei Province of China. Guo et al. [124] studied the incentive mechanism for enterprise's green innovation implementation.

Only five articles paid attention on the risk and innovation management for green innovation in China based on the literature review, like Xu and Gu [125] reviewed the risk management for enterprises green innovation and proposed a system risk management framework. Li [126] studied the industry transformation between the western and eastern of China; Chen and Xie [127] employed an Bayesian network model to monitor the innovation for green buildings; Fang [128] pointed that, the eco-culture development need to pay attention on the risk of eco-development; Li et al. [129] identified the risks in manufacturing industry and proposed the related suggestion as well. All these articles have a good reference for the practitioners of the green innovation execution in China [105].

35 articles implemented the theme of Innovation assessment, some the articles focus on the assessment method [130, 131], five articles discussed the innovation ability assessment, the representative authors are Sun and Cao [132], Xie et al. [133] and Jia [134]. Seven articles explored the innovation efficiency assessment like Huang et al. [135], Sun et al. [136], and Zhang et al. [137].

### 4.6 The role of entrepreneurship innovation

The last theme of green technology innovation is the role of entrepreneurship innovation. We divided this theme further into four sub-themes: entrepreneurship green design strategy, enterprise decision making and individual achievement. We caught five articles appeared from 1994 to 2019 involved with the role of individual. Some outstanding articles presented below.

Xu and Zhang [105], Jiang [138], found that green entrepreneur orientation has a positive impact on the environment and financial performance. Xie et al. [139] surveyed the resources enterprises' green behavior and found that benefit is one of the critical items for the green activities' participants.

Through above articles' exploration, it is known that that entrepreneurship is one of the key players in the process of green and sustainable innovation. However, the technological innovations used in the cooperation are mainly cost-oriented.

Our detailed exploration of these articles is not very extensive. However, we are witnessing green technology innovations involving many different types of topics that are more complex than our normal technological innovations because their impact has a profound impact on people's living environment and all different lines of business. And it is significant. Some topics have changed over time, leaving a lot of room for our researchers.

The themes of entrepreneurship dominates within the scope of the different type of entrepreneurs based on the previous review by Shi and Lai [2] are multiple such as the social corporate responsibility, the decision effect of entrepreneurs on enterprise's green technology innovation, and the role of entrepreneurs in green economy development or society improvement. While based on the review in China, it is not enough, noted that our selection of research is "green technology innovation", it is different. Moreover, the enterprise participant is crucial for green development, scholars are encouraging to pay attention on the role of entrepreneurship and education in cultivating entrepreneurs' green entrepreneurship, which is a critical element of sustainable development.

## 5. Result and discussion

### 5.1 Insight from our exploration

Through the analysis of the articles in CNKI from 1994 to 2019, we have identified that, although the research topics of green technology innovation are varied, the main stream is focusing on the technology adoption, diffusion, transfer, policy recommendation, or implementation, and advanced technology development currently, which can be regarded as convergence out of divergence.

Based on the overall review for the publication from 1994 to 2019, we caught out that green technology innovation is getting mature compared to the energy-innovation related topics. It is becoming a dominated research subject coupled with social energy system innovation contributing to the green and sustainable development.

In the period sequence analysis of the articles, we have initialized that China is acting as an emerging star and its publications obviously higher with the times goes, this is helpful for our researchers to have a full understanding of green development roadmap and trend in China.

Just seen from the analysis in Section 3.2, compared to the research methods of innovation, the articles with conceptual method account for 49%, and the qualitative method occupies 18%, both achieve 67%. It indicates that the green technology innovation in China is still at a developing stage and the method of model and empirical are not adequate. China's green development is still in the developing stage and need more executions. Though we know that methods should not be ignored with the extension and quick development. In addition, the actual technology adoption needs theoretical support. It assumes that the concepts and theoretical systems driving green technology development research is more welcome. Another impressive feature of these articles is that the green technology research covers 50 subjects within multi-disciplinary fields, while the majority falls on environment science, management, energy and fuels and economics.

The research level indicates that the green technology innovation in China is still under a developing stage based on the data shown in Section 3.3, herein, macroscopic research level gains 54%, mesoscopic research level are keeps 11%, microscopic research level are keeps 35%. This phenomenon reflects that more concrete studies and application from micro perspective are urgently needed, such as the specific technology innovation and application of new technology in the firms or environmental improvement.

The research subject area review in Section 3.4 shows that, the trend of green technology research appears an interdisciplinary research with the themes related to environmental subject, science technology, innovation management, which are accounting for 47.65% of the total. It is worth mentioning that the research articles of the top two areas–environmental sciences ecology and business economics are much more compared with other industries like agriculture and forest.

The keywords cluster analysis in Section 3.5, the technological innovation-related keywords gains highest occurrence frequency, such as "Innovation", "Technology", "Policy", "Sustainability", "System", "Performance", "Management", "Energy", and "Model" etc. The technological innovation is an important driving force for sustainable development. Various policies and adjusted industrial structures are introduced to promote green development of technologies, improve the efficiency of energy utilization, and support innovative development of green technologies, new and renewable energy, there are the key areas for our researcher to explore and develop.

For the review of publication based on institutions and publications in international journals in Sections 3.6 and 3.7, we can see that the most contributor in green technology innovation are science and technology university or organizations, especial for the universities from "985" and "211" project. Now China is executing a new program of Double First-Rate project[3] among the universities and reallocate the resources based on the contributions of the discipline construction and worldwide reputation. This including the national found support for the outstanding contributor or authors based on the evaluation system.

From the analysis in Section 4, one crucial phenomenon comes to our attention: besides the majority of research themes of "technology adoption and diffusion", green technology innovation cannot be isolated from the policy or regulation regime. The innovation research from multi-perspectives such as social culture,

[3] It refers the world first-class university and the first-class discipline construction.

economic management, engineering, social energy system and etc. are needed for low carbon development in China with green innovation orientation. This opens us a wider consideration that the green technology or is not just a pure technology but a cross functional activity, as a consequence, the technology becomes a low priority compared to the social innovation, such as the saving and efficiency improvement.

The review in Section 4.1 indicates that the effects of different policy implementations have different effects due to the different cost structure and maturity of technology. In fact, the green innovation can be induced by policies. The environmental regulation does the effectiveness on promoting the green technology development.

The finding in Section 4.2 shows: the technology adoption theory with empirical application based is also one of the key research themes. It covers innovative public procurement, dynamic efficiency analysis innovation activities and mechanisms design. People take great efforts on the real actual experiment and achievement in the adoption of green technology, such as the sample survey, case study or fields study with primary data. The typical research articles are influential. Green technology innovation diffusion takes leading the regional economic development. Furthermore, the researchers' focus is within new technology introduction, renewable energy development, new methods, new process improvement and even new conception or culture implantation. This is a new trend of green technology development in China.

In Section 4.3, the realization of technology transfer in green revolution is influenced by many factors. Green technology transfer needs to take full account of the capacity of absorbing countries and governments and the obstacles of intellectual property rights system. The increase of independent innovation capacity in China will play a positive role in technology transfer. From the perspective of historical development, in the environment of rapid improvement of independent innovation ability and intensified international competition, the transfer of green technology in developed countries will have a more positive impact on China's development, though China now is a second position in the world, many core technologies still need to import or transfer from the developed countries. It needs to increase the technology adaptation level and absorptive capacity to accept the technology transfer of the green revolution in a proactive manner.

While for the technological management and capability analysis in Section 4.4, it was sighted that the research of green technology innovation in enterprise management gradually turns from the macro aspect to the micro aspect, but it is not enough. After analyzing the green macro strategy of enterprises, scholars begin to pay more attention to micro factors such as enterprise organization, system design and innovation ability. Therefore, the government policies are becoming an important influencing factor of green innovation development, and the direction of future research.

In Section 4.5, the advanced technology development analysis has been explored from the concept, mechanism, technology foresight and multi-level perspective to enrich the related innovation theory. The basic research of policy instrument and mechanism among different interested group is necessary for technology innovation implementation. The basic research has covered the most areas of green technology innovation activities. It involves research of approaches policy instrument and mechanism, models, energy, environment resource industrial, biotechnology, etc. In addition, it is expected more interdisciplinary research to emerge in the future. While some publication level is not high and cannot get into the solid foundation to guide the green practice theoretically. Solving existing problems from different disciplines is also one of the directions of future research.

And at last, in Section 4.6, it was drawn that China need more entrepreneur and entrepreneurship. The scope of the different type of entrepreneurs are multiple such as the social corporate responsibility, the decision effect of entrepreneurs on enterprise's green technology innovation, and the role of entrepreneurs in green and low carbon economy development or society improvement. But the researches in China is not big, in fact, the enterprise participant is crucial for green development, scholars are encouraging to pay attention on the role of entrepreneurship and education in cultivating entrepreneurs' green entrepreneurship, which is a critical element for sustainability.

### 5.2 Recommendations for future direction

Based on the exploration of the content and trend analysis by different categories analysis from the perspectives of period sequence, research methods, research level, research subjects, keywords cluster, institution, authors and themes of articles in Sections 3 and 4, we have witnessed the researchers' contribution in China and are impressed with their consummate methodology and rich theory base. While some research limitations still can be caught due to the regional development differently in China. Herein, we summarize some of our recommendations of the future research directions for our researchers and practitioners as reference and discussion.

First, we would like to see our depth of understanding of green technology innovation and enrich our studies of methodology and theory. We recommend that researchers increase the research volume of theory extension and popularization of green life. Especial for the research from economics and social perspective, the approach in a concept innovation among human and enable of the studies getting deeper.

Second, we recommend our researchers greater use of experiments, field investigation and case study, which are the more solid and concrete foundations for the sustainability in China and would be more convincing for our practitioners.

Third, we propose the cross-level studies moving forward on the green technology promotion, such as the research of cross-culture among different enterprises, regions, organizations and other different stakeholders. The research does not only depend on the scientific research organizations, but mostly comes out of practice.

Fourth, green technology innovation is a global phenomenon with many countries serving as the locations where the technological innovation occurs. Thus, researchers or practitioners are encouraged to use cross-disciplinary teams worldwide to conduct truly international research, such as the global cooperation mechanism, global eco-system research, global cross-cultural studies on inter-action on the sustainable development etc. that means to implement the "Go Out" principles to keep the most advanced and updated research achievements.

Fifth, green technology innovation is a multi-level research and covers across different subjects and multidisciplinary subjects. The research with green technology is not only including the energy saving, renewable, sustainable consumption and transition research, technology development etc. but also including the human or social behavior research, eco-service, green accounting. Herein we also encourage researchers build research teams integrating science, environment, chemistry, energy, fuels, engineering, material science, social science, management, even psychology and other scientists from different fields, who can bring multiple perspectives and methodologies to the foundation of green innovation enhancement.

Finally, from the participant perspective of green technology innovation, government is the key for sustainability development, thus, the government policies are becoming an important influencing factor of green economy development. Moreover, as the main participants, the entrepreneur factors gained lots of scholars' attention as well, the themes of entrepreneurial spirit, corporate responsibility, effectiveness of entrepreneurs on enterprise's green technology innovation and the role of entrepreneurs becomes a new engine of green development in China.

### 5.3 Conclusion

This charter examined the literatures enlisted in the database of CNKI on the topics with regard to green technology innovation from 1994 to 2019 in China. Based on the literature review, some critical discussion and direction are drawn as follows: (1) green technology innovation is getting mature compared to the initial stage 10 years ago. Green technology is becoming a dominated research subject coupled with social energy system innovation contributing to the green and sustainable development. (2) The conceptual and qualitative publications dominate the overall researches; the empirical researches are in a shortage. (3) The research subjects are multi-perspective and multi-disciplinary, covering environment science, management, energy and fuels, economics and social behavior. New vibrancy of advanced theoretical and methodological research is particularly needed, especially for green technology innovation trajectory, performance evaluation, government policy instrument and multi-level cooperation among the participants. (4) The trend of green technology research appears an interdisciplinary research with the themes related to environmental subject, science technology, business economics, engineering and energy & fuels. (5) Different policy implementations have different effects due to the different cost structure and maturity of renewable energy.(6) Green technology innovation cannot be isolated from the policy or regulation regime, and is becoming a new underpin of current sustainable development coupled with social energy system contributing to eliminate the climate change.

From the most review of 25 years of the literatures within the domain of and green technology innovation it encourages us to conclude that the research is more diverse, more multi-faceted, multi-disciplinary and multi-focused than the normal technology innovation, but the publication amount is not big, most of the researchers in China still focus on the conceptual and qualitative exploration. Green innovation is a complex, multi-level and social constructed process that attract the researcher to perform in the developing fields. It shows strong evidence of the future trend on developing the new resource and renewable sources technology, new vibrancy of theoretical and methodological advance such as green technology innovation trajectory, innovation performance evaluation, government policy instrument and multi-level cooperation among enterprise, government policies etc. New and advanced theory explorations are the research themes of future directions.

## Acknowledgements

This research is supported by the Ministry of Education of Humanities and Social Science Fund (19YJA790037), Planning Project of Jiangxi Province (12th Five-year Plan) (No. 15GL28); the 9th China Postdoctoral Special Foundation (No. 2016T90789) and the Nature Science Foundation of Guangdong Province (No. 2018A030313269).

## Author details

Xiaodong Lai
School of Economic and Management, South China Normal University, Guangzhou, China

*Address all correspondence to: tylerlai@126.com

## References

[1] UK Energy White Paper: Our Energy Future - Creating a Low Carbon Economy, DTI (Department of Trade and Industry), Published by TSO (The Stationery Office) on 24 Feb 2003. Available from: www.tso.co.uk/bookshop

[2] Shi Q, Lai X. Identifying the underpin of green and low carbon technology innovation research: A literature review from 1994 to 2010. Technological Forecasting and Social Change. 2013;**80**(5):839-864

[3] Lv Y, Wang WQ. Research of enterprise' green technology innovation. Scientific Management Research. 1994; **12**(04):46-48

[4] Zhang JX, Zhu L. Research on technological innovation efficiency of industrial enterprises based on green growth of regions in China. The Journal of Quantitative & Technical Economics. 2012;**29**(02):113-125

[5] Xu SC, He ZX, Long RY. The effects of environmental regulations on enterprise green technology innovation. Science Research Management. 2012; **33**(06):67-74

[6] Zhang Q, Qu SY. Research on dynamic game between government and corporation environmental behavior and optimal strategies based on environmental regulation. Forecast. 2013;**32**(04):35-40

[7] Qian L, Xiao RQ, Chen ZW. Research on the industrial enterprise's technology efficiency and regional disparities in China-based on the theory of meta-frontier and DEA model. Economic Theory and Business Management. 2015;(01):26-43

[8] Luo LW, Liang SR. Green technology innovation efficiency and factor decomposition of China's industrial enterprises. China Population, Resources and Environment. 2016; **26**(09):149-157

[9] Li B, Peng X, Ouyang MK. Environmental regulation, green total factor productivity and the transformation of China's industrial development mode-analysis based on data of China's 36 industries. China Industrial Economics. 2013;**4**:56-68

[10] Lin RJ, Tan KH, Geng Y. Market demand, green product innovation, and firm performance: Evidence from Vietnam motorcycle industry. Journal of Cleaner Production. 2013;**40**:101-107. DOI: 10.1016/j.jclepro.2012.01.001

[11] Chen CF. China's industrial green total factor productivity and its determinants-an empirical study based on ML index and dynamic panel data model. Statistical Research. 2016; **33**(03):53-62

[12] Chen j, Liu JJ, Yang FM. Empirical research on the green technological innovation auditing. Studies in Science of Science. 2002;**1**:107-112

[13] Li WH, Bi KX, Sun B. Research on the effect environmental regulation intensity on green technological innovation of pollution intensive industries- empirical test based on panel data of 2003-2010. R&D Management. 2013;**25**(06):72-81

[14] Wan LL, Zhu Q. The impact of R&D input on the growth of industrial green total factor productivity-empirical data from China's industry from 1999 to 2010. Economic Perspectives. 2013;**9**:20-26

[15] Guo SQ. The theory and practice of ecological innovation of industry. Ecological Economy. 2002;**4**:34-37

[16] Dai HY, Liu XL. Some comment on the research of environmental

innovation. Studies in Science of Science. 2009;27(11):1601-1610

[17] Wu XB, Yang FM. The innovation and diffusion of green technology innovation. Science Research Management. 1996;**17**(01):38-41

[18] Chen J. Construction and analysis of national green technology innovation system. Studies in Science of Science. 1999;**17**(03):37-41

[19] Hua JY. Motivations and policy implications of low carbon technology innovation based on manufacturing industry. Science Research Management. 2011;**32**(06):42-48

[20] Qin SS, Yang S. An analysis of Xi Jinping's thought on green growth. Theory Journal. 2015;**6**:4-11

[21] Zhong H, Wang JF. Establish green technology innovation mechanism. Ecological Economy. 2003;**03**:41-44

[22] Xu QR, Wang Y. New observation on green technology innovation: Perspective form life cycle. Scientific Management Research. 1999;**17**(01):3-6

[23] Zhao XK. The influence of environmental policy on technology innovation. Journal of China University of Geosciences (Social Science Edition). 2004;**4**(1):24-28

[24] Dopfer K. The origins of meso economics. Journal of Evolutionary Economics. 2012;**22**(1):133-160

[25] Zhao B. Thoughts on the theory and practice of green economy. Social Science Research. 2006;**02**:46-47

[26] Zhou L. Spatial econometric analysis of green innovation in chin. Resources Science. 2010;**32**(05):932-939

[27] Li D, Yang JJ. Research on the characteristics and frontier of green technology innovation in China. Science Research Management. 2015;**36**(06): 109-118

[28] Lin SF, Sun J, Marinova D, Zhao DT. Evaluation of the green technology innovation efficiency of China's manufacturing industries: DEA window analysis with ideal window width. Technology Analysis & Strategic Management. 2018;**30**(10):1166-1181. DOI: 10.1080/09537325.2018.1457784

[29] Wang P, Wang ZH. Research on green technological innovation and development strategy for the enterprise in the pearl river delta. Urban Insight. 2011;**11**(04):134-140

[30] You QH, Wang P. Can environmental regulation promote R&D preferring to green technology research and development- based on the empirical research on China's industrial sector. Economic Review. 2016;**3**:26-38

[31] Wang FZ, Jiang T, Guo XC. Government quality, environmental regulation and enterprise green technology innovation. Scientific Research Management. 2018;**266**(01): 28-35

[32] Wang BC, Li HW. SEM model based empirical analysis on green product innovation affecting factors. China Population, Resources and Environment. 2009;**19**(5):168-174

[33] Zhang W, Li FL, et al. Theoretical model and thought discussion for the enhancement of China's green innovation ability through FDI. World Management. 2011;(12):170-171

[34] Xia C, Zhang LP. The evolutionary game analysis on green technological of enterprises under the environmental regulation –based on the view of stakeholder. Systems Engineering. 2017; **35**(02):103-108

[35] Li X. Optimal intertemporal decision-making of environmental

regulation under the background of supply-side. Science of Science and Management. 2017;**1**:46-53

[36] Zhang JX, Cai N, Mao JS, Yang C. Independent innovation, technology introduction and green growth of industry in China: An empirical research based on industry heterogeneity. Studies in Science of Science. 2015;**33**(02): 185-271

[37] Li Y. Impact of government policy and market competition on renewable energy innovation in EU countries. Resources Science. 2019;**41**(07): 1306-1316

[38] Li HJ, Zheng J, He YX. New energy automobile enterprise R&D investment and performance relationship. Forum on Science and Technology in China. 2017;**01**:76-93

[39] Cui XM, Yang DT, Liu CS. Research on the support system of green entrepreneurship based on the case study about photovoltaic- energy saving lighting, and new energy vehicles. Science and Technology Management Research. 2016;**36**(12):40-44

[40] Bi KX, Ding XH, Feng YJ. The present condition and develop trend of process innovation, measuring and appraise research in medium and small enterprises of manufacturing industry. Science Research Management. 2002; **06**:125-133

[41] Tian HN, Bi KX. Research on self-organization evolution of process greening innovation mechanism in manufacturing enterprises. Journal of Harbin Engineering University. 2012; **33**(09):1186-1193

[42] Ling HY, Song X, Guan CC. Innovative progress management helps green construction- on construction progress weight management based on BIM technology. Construction Enterprise Management. 2019;**04**:37-38

[43] Ren X, Zeng L, Zhou DD. Sustainable energy development and climate change in China. Climate Policy. 2005;**5**(2):185-198

[44] Jing D, Mingshan S. Integrated management for renewable energy resources and CDM resources: A case study. In: Globalization Challenge and Management Transformation. 2007

[45] Ou X, Zhang S, Chang S. Scenario analysis on alternative fuel/vehicle for China's future road transport: Life cycle energy demand and GHG emissions. Energy Policy. 2010;**10**(8):3943-3956

[46] Hu Z, Yuan J, Hu Z. Study on China's low carbon development in an economy-energy-electricity-environment framework. Energy Policy. 2011;**39**(5):2596-2605

[47] Chen S, Chen B. An optimization model for renewable energy generation and its application in China: A perspective of maximum utilization. Renewable and Sustainable Energy Reviews. 2012;**17**(9):1-5

[48] Ding W, Niu H, Chen J, et al. Influence of household biogas digester use on household energy consumption in a semi-arid rural region of Northwest China. Applied Energy. 2012;**12**(9): 16-23

[49] Liu H, Liang D. A review of clean energy innovation and technology transfer in China. Renewable & Sustainable Energy Reviews. 2013;**18**: 486-498

[50] Zhu X, Zhuang G, Xiong N. A review of China's approaches toward a sustainable energy future: The period since 1990. Wiley Interdisciplinary Reviews-Energy and Environment. 2014;**3**(5):409-423

[51] Shane SA, Ulrich KT. Technological innovation, product development, and entrepreneurship in management

science. Management Science. 2004; **50**(2):133-144

[52] Hou TS, Su ZD. Research on technological innovation effect of China's export industry chain resulted from green trade barrier. Studies in Science of Science. 2014;**04**:376-381

[53] Peng SZ, Sun XZ. Main challenges and strategic countermeasures for China's green economy development. China Population, Resources and Environment. 2014;**24**(03):1-4

[54] Feng ZJ, Chen W, Yang CJ. Environmental regulation difference, driven by innovation and china's economic green growth. Technology Economics. 2017;**36**(08):61-69

[55] Xu WH, Zheng JL. Environmental regulation boosts green technology innovation research: A test based on the Yangtze River economic belt. Contemporary Economics. 2019;**05**: 108-111

[56] Pei XJ, An X, Ye Y, Wang FY. Private investment, environmental regulation and green technological innovation–analysis of spatial Dubin model based on 11 provinces and cities in the Changjiang River economic belt. Science & Technology Progress and Policy. 2019;**36**(08):44-51

[57] Wu GZ, You DM. The influence mechanism of environmental regulation on technology innovation and green total factor productivity: Based on the moderating effect of fiscal decentralization. Journal of Industrial Engineering/Engineering Management. 2019;**33**(01):37-50

[58] Li WH. Spatial-temporal evolution and factors of industrial green technological innovation output in China's provinces: An empirical study of 30 provinces' data. Journal of Industrial Engineering Management. 2017;**31**(02): 9-19

[59] Gao GK, Wang YQ. Green innovation efficiency and its influencing factors of energy-intensive industry in Beijing-Tianjin-Hebei metropolitan region-empirical research based on spatial perspective. Journal of Industrial Technological Economics. 2018;**37**(01): 137-144

[60] Yang CJ, Yang WK, Zhao ZL. A research on influencing factors of regional green innovation in China. East China Economic Management. 2018; **32**(09):95-102

[61] Liu W, Xu JQ. Analysis on the affecting factors of strategic emerging industries' green technical innovation. Ecological Economy. 2018;**34**(11): 116-119

[62] Chen JX, Zhang YR. Research on regional green innovation efficiency in China and it's factors. Reformation & Strategy. 2018;**34**(06):72-79

[63] Li WH, Bi KX, Cao X. Impact of environmental regulation tools on green technological innovation in manufacturing enterprises': Taking papermaking and paper products manufacturing enterprises for example. Systems Engineering. 2013;**31**(10): 112-122

[64] Wang N, Qi YL, Dong H. The research on choices of policy tools in different environmental attitude situations. Science and Technology Management Research. 2016;**36**(16): 236-242

[65] Wei YR. Impact of tax policy on green transformation of manufacturing industry under green innovation. Reformation & Strategy. 2018;**34**(01): 98-128

[66] Shi CK. Effect and mechanism of market-based environmental tools on economic growth: Based on the perspective of industrial. Journal of

Central South University (Social Sciences). 2019;**25**(02):78-88

[67] Zhang J, Gen H, Xu GW, Chen J. Research of environmental regulation effectiveness on green technology innovation. China Population, Resources and Environment. 2019; **29**(01):168-176

[68] Li GP, Li YG, Quan JM. Environmental regulation, R&D investment and enterprises' green technological innovation capability. Science of Science and Management of S & T. 2018;**39**(11):61-73

[69] Xu JZ, Wang MM. Green technology innovation, environmental regulation and energy intensity-an empirical study based on Chinese manufacturing industry. Studies in Science of Science. 2018;**36**(04):744-753

[70] Francesco C, Claudia G, Francesco Q. Environmental and innovation policies for the evolution of green technologies: A survey and a test. Eurasian Business Review. 2015;**5**:2

[71] Wang HM. The impact of the internationalization of environmental regulation on china's foreign trade exports and countermeasures. Macroeconomic Management. 2010;**06**: 46-48

[72] Mannan S, Nordin SM, Rafik-Galea S, Rizal ARA. The ironies of new innovation and the sunset industry: Diffusion and adoption. Journal of Rural Studies. 2017;**55**:316-322

[73] Rennings K, Markewitz P, Vogele S. How clean is clean? Incremental versus radical technological change in coal-fired power plants. Journal of Evolutionary Economics. 2013;**23**(2): 331-355

[74] Sheng J, Chen MZ. Green food special vegetable innovation and efficiency technology. Jilin Vegetables. 2012;**07**:18

[75] Wan XX. Green construction innovation technology of Hong Kong-Zhuhai-Macao bridge. Informatization of China Construction. 2017;**08**:19-23

[76] Yan JL. Research on innovative technology application and energy conservation and environmental protection methods for green construction. Engineering and Technological Research. 2017;**03**:50-95

[77] Zhao XF. Innovative oil extraction technology to promote green sustainable development in the industry. China Petroleum and Chemical Industry. 2017; **02**:66-67

[78] Zhao F, Wang W, Li XY, Li HF, Han Y. Analysis of green energy saving innovation in Sunvalley, Shanghai Expo Park. Green Building. 2015;**7**(05):59-60

[79] Chai TX. Research of coal mining technology based on green theory. Energy and Energy Conservation. 2013; **12**:177-178

[80] Yun S, Lee J. Advancing societal readiness toward renewable energy system adoption with a socio-technical perspective. Technological Forecasting and Social Change. 2015;**95**:170-181. DOI: 10.1016/j.techfore.2015.01.016

[81] Zhou Y, Xu GN, Minshall T, Liu P. How do public demonstration projects promote green-manufacturing technologies? A case study from China. Sustainable Development. 2015;**23**(4): 217-231. DOI: 10.1002/sd.1589

[82] Zhu QH, Sarkis J, Lai KH. Green supply chain management innovation diffusion and its relationship to organizational improvement: An ecological modernization perspective. Journal of Engineering and Technology Management. 2012;**29**(1):168-185. DOI: 10.1016/j.jengtecman.2011.09.012

[83] Yan J. Research on green technological innovation of electronic appliance manufactures based on green purchasing. Scientific Management Research. 2010;**28**(5):102-105

[84] Liu P, Zhou DK, Yan JL, Zhou Y, Xue L. Promoting green manufacturing technology diffusion using an innovative policy implementation method. Engineering Sciences. 2016; **18**(04):101-108

[85] Fang ZQ. Research on the dispersion mechanism for green techniques innovation from the perspective of supply chain. Ecological Economy. 2018;**34**(06):63-67

[86] Chen SY, Chung CL. Exploring the relationships of green perceived value, the diffusion of innovations, and the technology acceptance model of green transportation. Transportation Journal. 2016;**55**:1

[87] Cao C, Wu XB, Zhou GG, Hu Q. Game analysis on innovation and diffusion of green product among manufacturers. Journal of Systems Engineering. 2012;**27**(05):617-625

[88] Liu HW, Liang X. Strategy for promoting low-carbon technology transfer to developing countries: The case of CCS. Energy Policy. 2011;**39**(6): 3106-3116

[89] Ai HS, Deng ZG, Yang XJ. The effect estimation and channel testing of the technological progress on China's regional environmental performance. Ecological Indicators. 2015;**51**:67-78

[90] Liang SR, Luo LW. International R&D capital technology spillover dynamic effects on green technology innovation efficiency. Scientific Research Management. 2019;**40**(03): 21-29

[91] Wu GC. The influence of green supply chain integration and environmental uncertainty on green innovation in Taiwan's IT industry. Supply Chain Management-An International Journal. 2013;**18**(5):539-552. DOI: 10.1108/scm-06-2012-0201

[92] Song YC, Yang WY. Research on the relationship between green technology innovation ability and financial performance based on factor analysis. Reform & Innovation. 2019;**07**:63-65

[93] Han JH, Yan SN. The impact of green technological innovation capability on value chain upgrade of manufacturing industry. Science and Technology Management Research. 2018;**38**(24):177-182

[94] Li JZ, Xiao YZ, Liu YN. Analysis on the influencing factors of green technological innovation ability of pulp and paper enterprise. China Forestry Economy. 2010;**05**:18-28

[95] Hua Z. Empirical study on china's eco-innovation capability and its influencing factors: Based on DEA-Malmquist productivity index. Technology Economics. 2011;**30**(09):36-69

[96] Liu ZS, Song DY, Liu GH. The threshold effect of environmental regulation on green technology innovation ability of manufacturing industry. Commercial Research. 2018; **04**:111-119

[97] Fu Y, Kok RAW, Dankbaar B, Ligthart PEM, van Riel ACR. Factors affecting sustainable process technology adoption: A systematic literature review. Journal of Cleaner Production. 2018;**205**:226-251. DOI: 10.1016/j.jclepro.2018.08.268

[98] Li Z. How to study the green development of economics in China. Reformation. 2016;(06):133-140

[99] Gao HG, Chen Z. Green economy in the perspective of two development outlooks. Ecological Economy. 2016;**32**:8

[100] Ma Y, Hou GS, Yin H. The evolution of technology innovation: From tradition to green. Science and Technology Management Research. 2014;**34**(19):11-15

[101] Zeng W, Qin SS. Study on construction of environmental supporting system about green consumption. Science and Technology Management Research. 2013;**33**(16): 233-236

[102] Lei G, Zhao X. Research on green innovation system in science and technology policy. Science and Technology Management Research. 2012;**32**(01):31-34

[103] Landsberger M, Onkham W, Cruz L, Rabelo L, Figueroa L. Hybrid real options for emerging technologies, green energy, and innovation. In: IIE Annual Conference. Proceedings. 2013

[104] Xu DL, Zhang ZJ. Research on the green & circular development policy system of the Qinba mountain area. Engineering Sciences. 2016;**18**(05): 74-79

[105] Xu JT, Zhang B. Study on the fiscal policy for promoting enterprises green technology innovation- from a perspective of the circular economy. Science and Technology Management Research. 2011;**31**(09):6-9

[106] He PP, Chen F. Reflections of the legislations on science and technology innovation under the background of ecological civilization. Journal of Hefei University of Technology (Social Science). 2016;**30**(05):57-61

[107] Su TJ. Research on the green culture construction of oil field-take Shengli oil field for example. Scientific Decision Making. 2010;**12**:42-52

[108] Li FM. Study on the path of leading ecological civilization construction with the concept of green development. China Collective Economy. 2017;**24**:1-2

[109] Guo YY, Zhang S, Zhang DP. The impacts of environmental regulation and governmental R&D funding on green technology innovation: Suppressing or promoting? East China Economic Management. 2018;**32**(07): 40-47

[110] Wang BB. Review of environmental policy and technological innovation. Economic Review. 2017; (04):131-148

[111] Li D, Yang JJ. A literature research on characteristics and trend for domestic green technology innovation. Science Research Management. 2015; **36**(06):109-118

[112] Narayanan VK. Research on technology standards: Accomplishment and challenges. Research Policy. 2012; **41**:1375-1406

[113] Wang P. Regional innovation milieus, green technology innovation and enterprise competitive advantage: A literature review. Forward Position or Economics. 2012;**3**(1):68-79

[114] Yang D, Chai HM. A review of driving factors of green technology innovation and its effect on firm's performance. China Population, Resources and Environment. 2015; **25**(11):132-136

[115] Zhang Q, Yao P. The impact of environmental regulation on the path and dynamic evolution of technological innovation based on porter hypothesis. Journal of Industrial Technological Economics. 2018;**37**(298):54-61

[116] Li L. Environmental regulation degree and enterprise green technology innovation performance-empirical studies based on structural equation model. Economic Forum. 2017;(04): 97-102

[117] Zhang X, Wang WW. Simulation of environmental regulation effects on green industrial development in China. Science-Technology and Management. 2018;**20**(01):34-44

[118] Zhen HZ. Research on the mechanism of green technology innovation and diffusion dynamics under regulatory conditions in Chinese enterprises. Scientific Management Research. 2018, 2016;**34**(5):77-88

[119] Liu YQ, Quan Q, Zhu J, Wang F. Green technology innovation, industrial agglomeration and ecological efficiency-a case study of urban agglomerations on Yangtze River economic belt. Resources and Environment in the Yangtze Basin. 2018;**27**(11):2395-2406

[120] Yang PY, Chen JJ. Research on industrial agglomeration, green development and governance system. Reform of Economic System. 2018;(05): 93-100

[121] Norberg-Bohm V. Stimulating 'green' technological innovation: An analysis of alternative policy mechanisms. Policy Sciences. 1999;**32**:1

[122] Yu FF. Green innovation behavior among housing developers: Impact of portfolio depth and scope. Science of Science and Management of S & T. 2015;**36**(05):13-23

[123] Zhang Z, Sun BW. The influence mechanism and empirical study on the efficiency of green growth in ecological economic belt of Han River in Hubei province: Based on the perspective of economic-social-environment-innovation subsystem. Ecological Economy. 2018;**34**(09):67-74

[124] Guo JD, Yang YQ, Ma ZC. Constraint and incentive mechanism for business green technology innovation. Science and Technology Management Research. 2018;**38**(20):249-252

[125] Xu XB, Gu JF. Research on the risk management of green technology innovation in enterprises. Modern Management Science. 2003;**12**:59-60

[126] Li B. Environmental risks and prevention and control measures to undertake industrial transfer in the central and western regions. Academic Journal of Zhongzhou. 2015;**10**:38-42

[127] Chen F, Xie HT. Risk analysis of construction technology innovation by Bayesian networks model-a case study of green building innovation project. Computer Engineering and Applications. 2014;**50**(18):33-38

[128] Fang SG. The construction of ecological civilization should be alert to the anti-ecological risk of technology. Gansu Theory Research. 2013;**04**:32-37

[129] Li HW, Zhang Q, Zhang YN. Manufacturing green product development risk identification research. Techno Economics & Management Research. 2011;**08**:29-32

[130] Bi KX, Zhu J, Feng YJ. On the index system to measure product innovation of small and medium enterprises. China Soft Science. 2002;**9**:56-59

[131] Zhu YY, Ma ZQ, Chen YQ. Multi-level fuzzy comprehensive evaluation on enterprise's green technology innovation environment. Science & Technology Progress and Policy. 2010; **27**(09):102-105

[132] Sun QY, Cao YK. An evaluation of enterprise green technology innovation ability based on extension theory. Science and Technology Management Research. 2016;**36**(21):62-67

[133] Xie JM, Tang XW, Shao YF. Research on stratified cluster evaluation of enterprise green technology innovation based on the rough set. Technology and Investment. 2012;**03**:02

[134] Jia J. Analysis of the correlation between foreign direct investment and the innovation ability of green technology in the host country. Science & Technology Progress and Policy. 2015;**32**(09):121-127

[135] Huang Q, Miao JJ, Li JY, Wang WH. Research on the spatial spillover effect of industrial enterprises' technology innovation efficiency based on green growth. Reform of Economic System. 2015;**04**:109-115

[136] Li YS, Cheng LM, Li Y. Ecological-economic efficiency evaluation of green technology innovation in strategic emerging industries based on entropy weighted TOPSIS method. Ecological Indicators. 2017;**73**:554-558

[137] Zhang HC, Li F, Zhang JP. Technological innovation efficiency evaluation of industrial enterprises in resource regions based on green growth perspective. Science and Technology Management Research. 2017;**37**(08): 69-76

[138] Jiang WB, Chai HQ, Shao J, Feng TW. Green entrepreneurial orientation for enhancing firm performance: A dynamic capability perspective. Journal of Cleaner Production. 2018;**198**:1311-1323. DOI: 10.1016/j.jclepro.2018.07.104

[139] Xie XB, Wu Y, Feng ZL, Hao ZT. Investigation of green behavior of resource-based enterprise in China. China Population, Resources and Environment. 2015;**25**(06):5-11

Chapter 10

# Politics of Renewable Energy in Africa: Nature, Prospects, and Challenges

*Samuel Adams and William Asante*

## Abstract

Energy transition discussions have centered on the technical, economic, and policy aspects of energy transitions. Despite this, the political dynamics have received less attention. It is suggested that since energy policy change threaten incumbent industries and impose substantial costs, enacting and sustaining policies require considerable political support. Even though it is widely acknowledged that barriers to energy transition are primarily political than technical, there is a lack of cohesive literature on the politics that drive, constrain, and shape renewable energy regulation or policy. This gap motivates this study. Adopting a desk research methodology and arguing from the lens of Kingdon's multiple streams framework, the study found among others that the streams of problem, politics and policy shows enough prospects to be coupled for Africa to make a serious consideration on its renewable energy capacity. However, a number of obstacles were also identified to make this venture difficult but are surmountable.

**Keywords:** renewable energy, multiple streams framework, Africa, politics, cleaner energy, energy transition, carbon dioxide emissions

## 1. Introduction

*"Alone we can do so little; together we can do so much"*
Quote attributed to Helen Keller, American Writer and Social Activist.

It is strongly believed that the United Nations (UN) member states were thinking just like Keller when they came up with the sustainable development goals (SDGs) as a means of combining their efforts to confront global developmental challenges of droughts, famine, poverty, climate change, and the likes. Following this, states are expected to "act locally" to ensure that the global SDG agenda is achieved by 2030. Since the discussion here focuses on renewable energy, rehashing Goal 7 of the SDGs is appropriate and situates this discussion in its proper context. The Goal 7 provides that member states should ensure access to affordable, reliable, sustainable and modern energy for all [1]. These include but not limited to ensuring universal access to affordable, reliable and modern energy services; increase substantially the share of renewable energy in the global energy mix; double the global rate of improvement in energy efficiency; enhance international cooperation to facilitate access to clean energy research and technology. The above targets consistently mention and encourage cleaner energy which makes the discussion on renewable energy timely and relevant.

Furthermore, the renewables agenda has become necessary because of global warming evidenced through storms and ice melts, droughts and hunger, unrest and migration [2]. This realization has led to a growing consensus directed at the transition to renewable energy systems, which has come to be known as a process of fuel substitution, a crucial way to addressing the climate crisis [2]. Renewable energy may refer to a form of energy that when used replaces itself and can last indefinitely when well-managed. The principal types of renewable energy consist of solar, thermal, photovoltaics, bioenergy, hydro, tidal, wind, wave, and geothermal [3].

The literature on renewable energy have mostly centered on three key academic fields i.e. political science, policy studies and energy transitions. However, Hughes and Lipsky [4] acknowledge that in political science the subfield of energy politics is "relatively underdeveloped." They continue that most of the studies fall within the 1970s and 1980s with prime focus on international political economy and oil geopolitics. It is in recent times that there is a gradual shift with new studies relating renewable energy to public opinion [5–7], electoral dynamics [8], coalitional politics [9], and green industrial constituencies [10]. Also, in the policy literature, scholars analyze renewable energy policymaking with theories such as the Multiple Streams Model, Punctuated Equilibrium theory and the Advocacy Coalition Framework. These studies emphasize windows of opportunity for policy change, especially following acute "focusing events" such as oil and nuclear crises [11–15]. The energy transition literature has centered on the technical, economic, and policy aspects of energy transitions [16, 17] with the political dynamics receiving less attention [4, 6]. It is suggested that since energy policy change threaten incumbent industries and impose substantial costs [18], enacting and sustaining policies require considerable political support. Even though it is widely acknowledged that barriers to energy transition are primarily political, there is a lack of cohesive literature on the politics that drive, constrain, and shape renewable energy policy particularly in developing countries [8, 19]. This study builds on the energy transition literature by exploring first, the energy situation in Africa, second, the prospects and challenges from transitioning from non-renewable to renewable, and third, the appropriate lessons that can be drawn to help Africa attain the SDG goal 7. The uniqueness of this study aside the above, is the application of Kingdon's multiple streams framework (MSF) as an analytical lens at the continental level.

By way of organization, the theoretical framework follows this discussion, followed by the methodology; then, the nature of energy situation in African is also discussed. After this, the discussion on whether conditions are ripe for the transition follows, and then a conclusion is drawn to end the chapter outlining some policy implications for the future.

## 2. Theoretical framework: multiple streams approach

The multiple streams framework (MSF) is acknowledged to be the handy work of John W. Kingdon, who explains how ideas come into being. In other words what makes important people pay attention to one subject rather than another, how their agendas change from time to time, and how they narrow their choices from a larger set of alternatives to very few [20–23]. These are the issues that the MSF seeks to explain. MSF views the policy process as composed of three streams of actors and processes. First, a problem stream consists of data about various problems

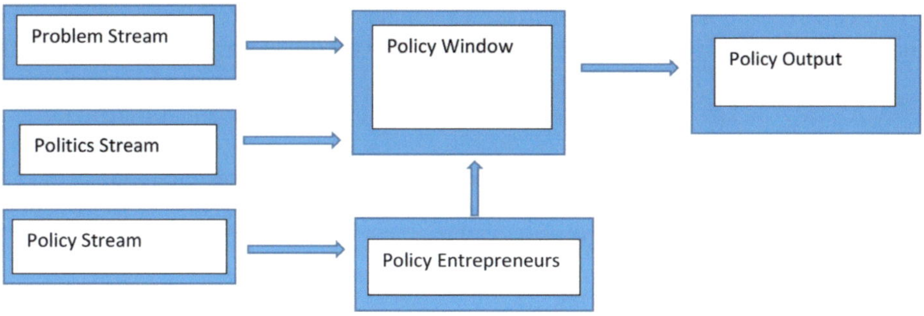


**Figure 1.**
*Multiple Streams Framework. Source: Zahariadis [25].*

and the proponents of various problem definitions. It may also consist of perceptions, opinions, and attitudes held by various members of the public and policy communities [20, 24]. Second, a policy stream which involves the proponents of solutions to policy problems that originate with communities of policy makers, experts and lobby groups. It is important to mention that the policy stream carries recommendations from researchers, advocates, analysts, who use their expertise to propose prospective solutions to them [24–26]. Third, a politics stream consists of elections and elected officials [20–23]. The politics stream also refers to factors such as changes in government, legislative turnover and fluctuations in public opinion. It must be mentioned that the political stream is often associated with contextual attributes such as the composition of ideas and values comprising national "moods" and the power shifts produced by legislative and executive turnover following events such as elections and cabinet shuffles that rotate the composition of policy-makers and affect important events through the composition of political and legislative timetables [24, 27].

According to Kingdon [21], the streams normally operate independently of each other, except when a "window of opportunity" permits policy entrepreneurs to couple the various streams. The success of the policy entrepreneurs in the coupling venture may result in a major policy change [20, 24, 27]. These policy entrepreneurs are vested stakeholders who strategically engage with the streams to open or seize windows of opportunities to advance their favored solutions [23]. In this framework, it is observed that policy development towards addressing a socio-economic problem does not occur automatically; rather, it emerges from the complex interaction and intersection of the three streams, which leads to certain issues being taken up by governments [24]. In this study we seek to explore the politics, problem and policy streams in the renewable energy sector of Africa and whether these streams are being strategically linked to enhance the acceptance of governments or policy makers. **Figure 1** illustrates the model.

## 3. Methodology

This study adopts a desk research methodology or desktop qualitative descriptive method. Desk research refers to the use of secondary data or that which can be collected without fieldwork. To most people it suggests published reports and statistics. In the context of this paper, the term is widened to include all sources of information that do not involve a field survey. These include searching libraries and the internet for data or information.

**Data Sources:** This paper benefitted extensively from text data on energy situation in SSA countries, renewable energy and its potential in Africa and other jurisdictions outside of Africa on the internet and in the libraries of University of Ghana and the Ghana Institute of Management and Public Administration (GIMPA).

**Data Analysis:** The information from the various sources of data were thoroughly studied and analyzed. Analysis was done with the objectives of the study at the background alongside the literature review.

## 4. The Africa energy situation

The Organization of African Unity (OAU) which is currently known as the African Union (AU) established the African Energy Commission (AFREC) through the Convention of the African Energy Commission (CAEC) adopted in Lusaka, Zambia, on 11 July 2001 and entered into force on 13 December 2006. After the Convention's adoption, it was expected that all member countries of the AU will be part of it, but as at January 14, 2019, only 35 countries had consented to its enforcement. Article 4 of the CAEC indicates that the AFREC is expected to map out energy development policies, strategies and plans based on sub-regional, regional and continental development priorities and recommend their implementation in member countries. This energy commission architecture is hoped to propel Africa's energy situation to its peak but the continent still records the lowest share in terms of access to power by its citizens. World Bank [28] records that the percentage of SSA population with access to electricity is pegged at approximately 44.6%, which suggests that a lot still needs to be done for the people of Africa. It is observed that Africa's energy sector is dominated by fossil fuels, hydro, nuclear and biomass (see **Table 1**).

Further, it must be pointed out that traditional biomass energy use (wood, charcoal, agricultural residues and animal waste) and fossil fuels contribute to respiratory illnesses in highland areas of sub-Saharan Africa because of the excess $CO_2$ emissions [30]. This however calls for the need to look at other sources of clean energy supply. Tables below show the renewable energy potential of African countries (see **Tables 2** and **3**). Additionally, various types of renewable energy have been identified to exist and can be tapped by African countries, these include geothermal, hydropower, wind energy, solar and bioenergy. It must be pointed out that some countries have already taken the lead in tapping these forms of renewable energy (see **Table 4**).

| Category | 2000 | 2005 | 2010 | 2015 |
|---|---|---|---|---|
| Production of electricity from biofuels and waste | 135 | 163 | 187 | 349 |
| Production of electricity from fossil fuels | 29,921 | 37,321 | 44,975 | 62,212 |
| Production of nuclear electricity | 1,119 | 971 | 1,101 | 1,221 |
| Production of hydro electricity | 6,607 | 8,107 | 9,738 | 12,495 |
| Production of geothermal electricity | 37 | 77 | 126 | 329 |
| Production of electricity from solar, wind, etc. | 20 | 128 | 326 | 1,086 |
| *Source: Extracted from AFREC [29].* | | | | |

**Table 1.**
*Summary of Africa's Total Energy Production (Ktoe) 2000–2015.*

| Country | 2009 (MW) | 2018 (MW) |
|---|---|---|
| Congo DR | 2514 | 2750 |
| Egypt | 3354 | 4813 |
| Ethiopia | 1443 | 4326 |
| Ghana | 1187 | 1659 |
| Morocco | 1520 | 3263 |
| Mozambique | 2198 | 2235 |
| Nigeria | 2087 | 2143 |
| Sudan | 1681 | 2136 |
| Zambia | 1723 | 2446 |

*Source: Extracted from IRENA [31].*

**Table 2.**
*Renewable energy capacity (MW) (leading African countries).*

| Country | 2009 (GWh) | 2017 (GWh) |
|---|---|---|
| Angola | 3308 | 7897 |
| Cameroon | 4017 | 5106 |
| Congo DR | 7940 | 9287 |
| Cote D'Ivoire | 2132 | 2054 |
| Egypt | 15942 | 15957 |
| Ethiopia | 3593 | 12585 |
| Ghana | 6893 | 5672 |
| Kenya | 3923 | 8407 |
| Malawi | 1813 | 1915 |
| Morocco | 2976 | 4706 |
| Mozambique | 16994 | 14127 |
| Namibia | 1405 | 1526 |
| Nigeria | 7454 | 7803 |
| South Africa | 1648 | 10453 |
| Sudan | 3379 | 9484 |
| Tanzania | 2738 | 2611 |
| Uganda | 1458 | 3745 |
| Zambia | 10604 | 12537 |
| Zimbabwe | 5517 | 4214 |

*Source: Extracted from IRENA [31].*

**Table 3.**
*Renewable energy production (GWh) (leading African countries).*

## 5. Findings and discussion

This section of the chapter discusses the MSF perspective with respect to the identified constructs of the framework; problem stream, policy stream, politics stream, policy entrepreneurs and policy window. This is followed by an attempt

| Renewable energy | Leading African countries with capacity (MW) | Leading African countries in production (GWh) |
|---|---|---|
| Geothermal | Ethiopia, Kenya | Kenya |
| Hydropower | Congo DR, Egypt, Ethiopia, Ghana, Morocco, Mozambique, Nigeria, South Africa, Sudan, Zambia | Angola, Cameroon, Congo DR, Cote D'Ivoire, Egypt, Ethiopia, Ghana, Kenya, Malawi, Morocco, Mozambique, Namibia, Nigeria, South Africa, Sudan, Tanzania, Uganda, Zambia, Zimbabwe. |
| Wind | Egypt, Ethiopia, Morocco, South Africa, Tunisia. | Egypt, Ethiopia, Morocco, South Africa, Tunisia. |
| Solar | Algeria, Egypt, Morocco, Reunion, South Africa | Algeria, Egypt, Morocco, Reunion, South Africa |
| Bioenergy | Eswatini, Ethiopia, South Africa, Sudan, Zimbabwe | Angola, Egypt, Eswatini, Kenya, Mauritius, Reunion, South Africa, Sudan, Tanzania, Uganda, Zimbabwe |

*Source: Extracted from IRENA [31].*

**Table 4.**
*Types of renewable energy capacity and production in Africa.*

to discuss the extent to which the three streams are being coupled for an effective energy transition on the African continent.

## 5.1 Problem stream

Proponents of the MSF are of the view that for a policy to be considered, it should be politically and technically feasible. Additionally, its capacity to address the corresponding problem that it is expected to address should be relevant. MSF Proponents have pointed out clearly that the problem stream may refer to policy problems in society that potentially require attention [22–25]. In the case of this study, it is observed that Africa's economy is growing at unprecedented rate, and one of the core challenges associated with this economic growth is related to energy constraints. Specifically, economic growth, changing lifestyles and the need for reliable modern energy access require energy supply to be at least doubled by 2030 [32]. An investment of about US$43–55 billion per year is needed until 2030–2040 to meet demand and provide universal access to electricity. However, the present investment situation in the energy sector is about US$8–9.2 billion which is woefully inadequate [33]. Furthermore, it is estimated that over 645 million people do not have access to electricity. Again, out of the world's 20 countries with the least access to electricity, 13 can be found in Africa, including Nigeria, Ethiopia, Democratic Republic of the Congo (DRC), Tanzania, Kenya, Uganda, (the former) Sudan, Mozambique, Madagascar, Niger, Malawi, Burkina Faso, and Angola [33]. This has however heightened the need to focus on tapping the enormous renewable energy potential in Africa that is untapped [33].

Again, energy security is a socio-economic and political factor that contributes to sustainable development (SD) in any nation. Currently, the world is dominated by the usage of non-renewable energy such as fossil fuels. The use of non-renewable energy leads to the emission of large amounts of greenhouse gases (GHGs), which is considered to be the principal cause of climate change. Accordingly, the use of clean energy sources to reduce the release of carbon emissions is a key goal in reducing global warming and promoting sustainable development [33–36].

### 5.2 Policy stream

The policy stream in the view of MSF proponents pertains to the many potential policy solutions that originate with communities of policy makers, experts and lobby groups. This may also include recommendations from researchers, advocates, analysts, and others in a policy community examining problems and using their (sometimes self-proclaimed) expertise to propose prospective solutions to them [24, 37, 38]. Many international energy organizations and researchers have proposed tapping the untapped African renewable resources in confronting the continent's energy challenge [32, 33, 36]. **Table 2** shows African countries that have renewable capacity of 1000 MW and above. Additionally, **Table 4** shows five key renewable energy sources such as geothermal, hydropower, wind, solar and bioenergy alongside corresponding African countries with capacity to be tapped. It is however important to state that some African countries have already begun this effort but more attention should be geared toward that direction.

### 5.3 Politics stream

The third of the streams according to proponents of the MSF is the politics stream. Here, emphasis is placed on factors such as changes in government, legislative turnover following events such as elections and cabinet shuffles that rotate the composition of policy makers and fluctuations in public opinion [24, 25]. Although efforts were made in the early 2000s at the continental level with the adoption of the Convention of African Energy Commission, the renewables agenda has gained momentum in recent times due to the adoption of the AU Agenda 2063. This agenda is a shared framework for inclusive growth and sustainable development for Africa to be realized in the next fifty years. A new crop of African leaders have realized that African problems can only be solved by Africans, and so they agreed in 2013 through the 50th Anniversary Solemn Declaration during the commemoration of the Fiftieth Anniversary of the Organization of African Unity (OAU) to bring forth the Agenda 2063 which consist of seven aspirations [39]. This is to guide individual member states in their own development planning. This has also led to some reforms in the AU especially the shift to self-financing of the AU policies and programs [40].

### 5.4 Policy entrepreneurs

Policy entrepreneurs are vested stakeholders who strategically engage with the streams to open or seize windows of opportunities to advance their favored solutions [23]. The literature reveals that various bodies within member countries are in charge of energy in general and renewable energy to be specific [30, 41]. Aside the AFREC which is supposed to carry out research activities and inform policy directions of AU member states in the energy sector, many other international institutions and NGOs either directly or indirectly influence Africa's energy governance with their activities. Some of these actors include the International Energy Agency (IEA), the Intergovernmental Panel on Climate Change (IPCC), the United Nations Environment Programme (UNEP), the United Nations Framework Convention on Climate Change (UNFCCC), the World Bank, the European Renewable Energy Council (EREC), the OPEC, African Development Bank, International Renewable Energy Agency (IRENA), the Renewable Energy and Energy Efficiency Partnership (REEEP) and the Sustainable Energy for All (SE4ALL) [42]. Some of the strategies adopted by these actors include for example the World Bank and AfDB may target their lending activities to projects related to energy or renewable energy. The IRENA on the other hand may adopt innovative strategies to promote renewable energy by concentrating on a narrowly defined set of goals with regards to the deployment of renewables and additionally provide epistemic services as well [42]. The study shows that the policy entrepreneurs to push the renewable energy agenda are vast and wide, with some of them already in the process (WB, AfDB, IRENA, UNEP, and AFREC).

### 5.5 Policy windows

MSF proponents are of the view that policy windows present opportunities that pave the way for policy entrepreneurs to push their policy ideas to the policy makers/governments [24]. The AU Agenda 2063 is one of the continental policy windows with regards to the adoption of a renewable energy policy. The first aspiration of Agenda 2063 posits "a prosperous Africa based on inclusive growth and sustainable development" [39]. It could be argued that this aspiration opens doors to discuss issues of renewable energy which could aid in inclusive growth and sustainable development. Another important policy window is the rise in population growth which has shifted focus on alternative sources of energy. Again, in recent times the increase in oil prices and the sustainable development Goals has further heightened interests in the call for renewable energy. As stated earlier, Goal 7 enjoins all UN member states to ensure access to affordable, reliable, sustainable and modern energy for all [1].

Aside the above policy windows, two important global environment initiatives have also stimulated greater interest in renewables in Africa. The first was the United Nations Conference on Environment and Development (UNCED) held in Rio de Janeiro, Brazil in 1992. At this Conference, an ambitious environment and development document entitled "Agenda 21" was reviewed by one of the largest gathering of Government Heads of States and endorsed by a large number of multinationals companies. Agenda 21 sought to operationalize the concept of sustainable development. In addition, the Rio Conference provided the venue for the second important event, the signing of the United Nations Framework Convention on Climate Change (UNFCCC) by 155 Governments. The Convention came into force in early 1994 after ratification by 50 States. Renewables featured in both Agenda 21 and the Climate Change Convention. In addition, renewables featured high on the agenda of the Johannesburg World Summit on Sustainable Development (WSSD) in 2002. In the UN-led implementation plan of action for the WSSD, dubbed WEHAB (which stands for Water, Energy, Health, Agriculture and Biodiversity), top priority was given to the renewables and other alternative forms of energy services. One of the targets proposed at WSSD was for every country to commit itself to meeting 10% of its national energy supply from renewables [30].

### 5.6 Coupling the streams and the associated challenges

The multiple streams framework suggests that the ability of a policy entrepreneur(s) to strategically couple the streams of problem, policy and politics through a window of opportunity and with the consent of policy makers, gives a high possibility for a policy to be adopted or an issue to get to the agenda stage [20–25]. The ensuing discussion clearly reveals that all the necessary factors are in place as proposed by the MSF. The question to ask then is: Has coupling of the streams been successful on the continental level? The discussions above points to the fact that prospects are high especially when you want to focus on Goal 7 of the SDGs, that proposes that member states should ensure access to affordable, reliable, sustainable and modern energy for all [1]. The contribution of Africa to ensure universal access to affordable, reliable and modern energy services, to increase substantially the share of renewable energy in the global energy mix and to enhance international cooperation to facilitate access to clean energy research and technology, including renewable energy are yet to be realized, as it is recorded that the percentage of Sub-Saharan African population access to electricity is still below average [28] and then the existence of substantial potential of renewable energy resources that also remain untapped [31, 33]. It must be quickly pointed out that these are early days yet as the SDGs were born barely 4 years ago and the AU Agenda 2063 also about 6 years when it was agreed upon. This suggests that Africa has not been idle and it will be unfair on our part to gloss over the modest efforts being pursued by member countries of the AU.

However, despite the above, it is appropriate to focus attention on the factors that are likely to work against the efforts of African countries. First, it is observed that African countries differ in a number of ways, for example institutional frameworks and governance systems differ greatly. Some have open systems whereas others still practice a closed system. According to Gordon [43], the Ethiopian state is tightly controlled by the ruling coalition, the Ethiopian People's Revolutionary Democratic Front (EPRDF). The EPRDF is made up of four constituent parties based on ethnic groups, consisting of the Tigrayan People's Liberation Front (TPLF), the Amhara National Democratic Movement (ANDM), the Oromo Peoples' Democratic Organization (OPDO), and the Southern Ethiopian People's Democratic Movement (SEPDM). Ethiopia is regarded as a country with a very high public investment rate but a low private investment rate. As a result the largest companies in the country are state-owned; those found to be private are owned by close allies of powerful political elites. On the other hand, in Kenya for example, private companies have been present for decades and the country has become a hub for innovation in commercial off-grid and micro-grid systems [43]. These experiences reflect different political, regulatory, and security environment and therefore poses contextual challenges to push for a collective renewable agenda without a comprehensive assessment and understanding of AU member states.

The legislative and regulatory constraints in many African countries make it difficult to embark on a sustainable energy policy that would be workable in national jurisdictions. Different states have their own strategies in dealing with similar issues. And so a one size fits all renewable energy strategy will not suffice. Again, electoral related conflicts and other forms of ethnic-based violence in places like Congo DR, Sudan, Cote D'Ivoire, Kenya, Somalia, etc. poses security risks. Gordon [43] reports that the risk of protests represents the greatest physical threat to renewable energy assets. He recounts that between 2015 and 2018, protests recurred in Oromia, Amhara, and to a lesser extent in Addis Ababa and the Somali region. Protests often attracted thousands of people, and in Oromia and Amhara led to attacks on foreign businesses, particularly those that were either associated with the ruling party or those that were central to the government's economic policy. Additional challenge is the overreliance on international financing. This comes with a lot of conditionalities that are sometimes unfavorable to the socio-economic and political environment of most African countries. This calls for carefulness and due diligence in international business transactions but at the same time to reduce undue delays in contract performance. A classic case is the Lake Turkana Wind Power Project in Kenya which took nine years to reach financial closure. There are other ongoing projects like the Corbetti Geothermal project in Ethiopia which has taken 7 years and still counting [43].

Also, the issue of infrastructure and skilled human resources leave much to be desired in this venture. It is observed that majority of infrastructure projects on the continent were financed by large international agencies because of the large costs involved. Also, the unstable economic environments highlighted in currency exchange rate fluctuations, depreciation and high interest rates do not provide opportunities for indigenous businesses to thrive.

Additionally, inadequate planning policies, lack of co-ordination and linkage in the Renewable Energy Technology program, weak dissemination strategies, poor baseline information and, weak maintenance service and infrastructure [30] are crucial challenges that should not be ignored. It is important to state that there is always an advantage in cooperation, and so the prospects to consider renewable energy as a very important energy source in Africa's energy mix should be sustained at worst and intensified at best if Africa is to make any meaningful contribution to the SDGs and its own Agenda 2063.

## 6. Conclusion and policy implications

This study has added to the extant literature on energy transition by exploring the situation in Africa and how best the continent can increase its energy mix with a focus on renewable energy. The study has clearly demonstrated that the energy situation in Africa has been dominated by fossil fuels which cause excessive emission of $CO_2$ in to the atmosphere leading to climate change with its attendant consequences. A number of development opportunities have opened the door for discussions on renewable energy in Africa and the world at large. Excerpts of these opportunities include the famous SDGs, the AU Agenda 2063, the Earth Summit in Rio, the World Summit on Sustainable Development in Johannesburg, South Africa in the year 2002 and other important multilateral and bilateral treaties. Again, the study has revealed that the dominant renewable energy resources on the continent include geothermal, hydropower, wind energy, solar and bioenergy; and that many countries have already taken the lead in exploiting these resources [31].

Moreover, the study has shown that from the MSF perspective, the prospects of transitioning to renewable energy is high, in the sense that, the problem has been clearly defined as the huge gap in energy supply to the growing African population, the cost and scarcity associated with fossil fuels, and the excessive emission of greenhouse gases into the atmosphere leading to climatic changes by fossil fuels. African political elites and international actors have all agreed that renewable energy is not a bad idea and can add significantly to the energy mix in Africa. However, the challenge of political insecurity, infrastructural inadequacies, confusion in regulatory and institutional frameworks in some African countries, lack of skilled manpower, initial cost and investment risks are but a few of these challenges that require enormous attention if progressed is to be witnessed in the renewable energy venture.

Furthermore, in order to make significant progress in attaining the renewable energy objective as reflected in SDG Goal 7 by African countries, five main policy implications are worth considering. First, there should be conscious efforts of planning and developing a comprehensive renewable energy implementation framework for African countries with AFREC providing a supervisory role to ensure that AU member states are prioritizing renewable energy considerations in their domestic energy policies and programs. Second, efforts should be made to carefully select renewable energy technologies that are conducive to the African environment. In doing this, there should be a deliberate attempt by governments to train Africans in various aspects of renewable energy technologies. Third, the economic environment should be made conducive to attract and support indigenous Africans who are interested and want to embark on renewable energy investments.

Fourth, governments should take it upon themselves to ensure that regulatory and institutional frameworks are harmonized, made simple, easily accessible and easily understood to clearly define the parameters for investors, governments and the citizens/indigenous people. Fifth, AU member states can and should explore the option of trading electricity among themselves to widen the market base for renewables and curb the challenge of intermittent power supply. In conclusion, renewable energy comes along with many prospects in promoting national development; as a result, serious attention should be given to it at the national, sub-regional and continental levels.

## Abbreviations

| | |
|---|---|
| AfDB | African development bank – An international Bank for African countries that aims to encourage sustainable economic development and social progress thus contributing to poverty reduction. |
| AFREC | African energy commission – This body was set up by the Organization of African Unity (OAU) now known as AU to map out energy development policies, strategies and plans based on sub-regional, regional and continental development priorities and their implementation. |
| AREF | African renewable energy fund – This is a fund established to support small to medium scale independent power producers (IPPs) across sub-Saharan Africa. It is managed by Berkeley Energy, an experienced renewable energy fund manager active in Asian and African emerging markets. |
| AU | African Union – An intergovernmental organization with African Countries as members. It was formed in 2002 for mutual cooperation to replace the defunct Organization of African Unity. |
| EREC | European renewable energy council – This was created in the year 2000, and it is the umbrella organization of the major European renewable energy industry, trade and research associations active in the field of photovoltaics, small hydropower, solar thermal, geothermal, etc. |
| GWh | Giga Watts per hour – A unit of energy representing one billion watt hours and equivalent to one million kilowatts hour. It is used as a measure of the output of large electricity power stations. |
| IEA | International Energy Agency – An intergovernmental organization that ensures reliable, affordable and clean energy for its 30 member countries and beyond. Their mission is focused on 4 main areas: energy security, economic development, environmental awareness and engagement worldwide. |
| InfraCo Africa | InfraCo Africa is part of the multilateral Private Infrastructure Development Group (PIDG). InfraCo Africa seeks to alleviate poverty by mobilizing private investment into infrastructure projects in sub-Saharan Africa's poorest countries to the highest standards. |
| IPCC | Intergovernmental Panel on Climate Change – This is the UN body for assessing the science related to climate change. It provides regular assessments of the scientific basis of climate change, its impacts and future risk and options for adaptation and mitigation. |
| IRENA | International Renewable Energy Agency – An intergovernmental organization that supports countries in their transition to a sustainable energy and serves as the principal platform for |

| | |
|---|---|
| | international co-operation, a center of excellence, and a repository of policy, technology, resource and financial knowledge on renewable energy. |
| Ktoe | Kiloton of Oil Equivalent – This is a unit to measure the amount of Energy released by burning a thousand tonnes of crude oil. |
| MSF | Multiple streams framework – A theory developed by John Kingdon to explain agenda setting in the policy making process. |
| MW | Mega Watt – This is a unit of power equal to one million watts, especially as a measure of the output of a power station. |
| PIGD | Private Infrastructure Development Group – It encourages and mobilizes private investment in infrastructure in the frontier markets of sub-Saharan Africa, south and south-east Asia, to help promote economic development and combat poverty. Since 2002, PIDG has supported 154 infrastructure projects to financial close and provided 222 million people with access to new or improved infrastructure. |
| REEEP | Renewable energy and energy efficiency partnership – A body that develops innovations, efficient financing mechanisms to advance market readiness for clean energy services in low and middle-income countries. |
| SE4ALL | Sustainable energy for all – An independent not for profit international organization with headquarters in Vienna, Austria. Some of its priority areas include electricity for all in Africa, energy for displaced people, energy and health, etc. |
| UNEP | United Nations Environment Programme – It is part of the UN system. It is the arm of the UN that takes charge of all environment-related issues. |
| UNFCCC | United Nations Framework Convention on Climate Change – It is part of the UN system, and established in 1992. It is tasked with supporting the global response to the threat of climate change. |

## Author details

Samuel Adams* and William Asante
School of Public Service and Governance, Ghana Institute of Management and Public Administration, Accra, Ghana

*Address all correspondence to: sadamss2000@yahoo.com

## References

[1] United Nations Economic Commission for Africa. Accelerating SDG 7 Achievement Policy Briefs in Support of the First SDG 7 Review at the un High-Level Political Forum 2018. Retrieved from: https://sustainabledevelopment.un.org/content/documents/18041SDG7_Policy_Brief.pdf

[2] Burke MJ, Stephens JC. Political power and renewable energy futures: A critical review. Energy Research & Social Science. 2018;**35**:78-93. DOI: 10.1016/j.erss.2017.10.018

[3] Boyle G. Renewable Energy. Oxford: Oxford University Press; 2004

[4] Hughes L, Lipscy PY. The politics of energy. Annual Review of Political Science. 2013;**16**:449-469

[5] Ansolabehere S, Konisky DM. Cheap and Clean: How Americans Think About Energy in the Age of Global Warming. Cambridge, Massachusetts: MIT Press; 2014

[6] Stokes LC. The politics of renewable energy policies: The case of feed-in tariffs in Ontario, Canada. Energy Policy. 2013;**56**:490-500. DOI: 10.1016/j.enpol.2013.01.009

[7] Stokes LC, Warshaw C. Renewable energy policy design and framing influence public support in the United States. Nature Energy. 2017;**2**(17107):1-6. DOI: 10.1038/nenergy.2017.107

[8] Stokes LC, Breetz HL. Politics in the US energy transition: Case studies of solar, wind, biofuels and electric vehicles policy. Energy Policy. 2018;**113**:1-43. Available at: https://escholarship.org/uc/item/8v19r289

[9] Meckling J, Jenner S. Varieties of market-based policy: Instrument choice in climate policy. Environmental Politics. 2016;**25**(5):853-874. DOI: 10.1080/09644016.2016.1168062

[10] Aklin M, Urpelainen J. Political competition, path dependence, and the strategy of sustainable energy transitions. American Journal of Political Science. 2013;**57**(3):643-658. DOI: 10.1111/ajps.12002

[11] Carlisle JE, Feezell JT, Michaud KE, Smith ER. The Politics of Energy Crises. Oxford: Oxford University Press; 2016

[12] Grossman PZ. US Energy Policy and the Pursuit of Failure. Cambridge: Cambridge University Press; 2013

[13] Nohrstedt D, Weible CM. The logic of policy change after crisis: Proximity and subsystem interaction. Risk, Hazards and Crisis in Public Policy. 2010;**1**(2):1-32

[14] Nohrstedt D. The politics of crisis policymaking: Chernobyl and Swedish nuclear energy policy. Policy Studies Journal. 2008;**36**(2):257-278

[15] Smith ERAN. Energy, the Environment, and Public Opinion. Lanham: Rowman and Littlefield Publishers; 2002

[16] Smil V. Energy Transitions: History, Requirements, Prospects. Santa Barbara: Praeger; 2010

[17] Schwerhoff G, Sy M. Developing Africa's energy mix. Climate Policy. 2019;**19**(1):108-124

[18] Breetz HL, Mildenberger M, Stokes LC. The political logics of clean energy transitions. Business and Politics. 2018;**20**(4):492-522

[19] Delucchi MA, Jacobson MZ. Providing all global energy with wind, water, and solar power. Part II:

Reliability, system and transmission costs, and policies. Energy Policy. 2011;**39**(3):1170-1190

[20] Kingdon JW. Agendas, Alternatives, and Public Policies. 2nd ed. New York: Harper; 1995

[21] Kingdon JW. Agendas, Alternatives and Public Policies. 2nd ed. England: Pearson Education Limited; 2014

[22] Howlett M, McConnell A, Perl A. Streams and stages: Reconciling kingdon and policy process theory. European Journal of Political Research. 2015;**54**:419-434

[23] Ruseva T, Foster M, Arnold G, Siddiki S, York A, Pudney R, et al. Applying policy process theories to environmental governance research: Themes and new directions. Policy Studies Journal. 2019;**47**(S1):S66-S95

[24] Howlett M, McConnell A, Perl A. Moving policy theory forward: Connecting multiple stream and advocacy coalition frameworks to policy cycle models of analysis. Australian Journal of Public Administration. 2016;**76**(1):65-79

[25] Zahariadis N. The multiple streams framework: Structure, limitations, prospects. In: Sabatier P, editor. Theories of the Policy Process. 2nd ed. Boulder, CO: Westview Press; 2007

[26] Voß J-P, Simons A. Instrument constituencies and the supply side of policy innovation: The social life of emissions trading. Environmental Politics. 2014;**23**(5):735-754

[27] Stimson JA. Public Opinion in America: Moods Cycles and Swings. Boulder: Westview Press; 1991

[28] World Bank. Sustainable Energy for All (SE4ALL) Database from the SE4ALL Global Tracking Framework led jointly by the World Bank, International Energy Agency, and the Energy Sector Management Assistance Program. 2017. Retrieved from https://data.worldbank.org/indicator/EG.ELC.ACCS.ZS?locations=ZG

[29] AFREC. AFREC Africa Energy Database. Algiers: African Energy Commission (AFREC); 2015

[30] Karekezi S, Kithyoma W. Renewable energy development. In: The Workshop for African Energy Experts on Operationalizing the NEPAD Energy Initiative; 2-4 June 2003. Senegal: Novotel, Dakar; 2003

[31] IRENA. Renewable Energy Statistics 2019. Abu Dhabi: The International Renewable Energy Agency; 2019

[32] IRENA. Africa 2030: Roadmap for a Renewable Energy Future. IRENA: Abu Dhabi; 2015

[33] UNEP. Atlas of Africa Energy Resources. Kenya: UNEP; 2017

[34] Owusu PA, Asumadu-Sarkodie S. A review of renewable energy sources, sustainability issues and climate change mitigation. Cogent Engineering. 2016;**3**:1-15

[35] International Energy Agency. Energy Climate and Change World. Energy Outlook Special Report. 2015. Retrieved from: https://www.iea.org/publications/freepublications/publication/WEO2015SpecialReportonEnergyandClimateChange.pdf

[36] Souza GM, Ballester MVR, de Brito Cruz CH, Chum H, Dale B, Dale VH. The role of bioenergy in a climate-changing world. Environmental Development. 2017;**23**:57-64

[37] Lauber V. Political economy of renewable energy. In: Wright JD, editor. International Encyclopedia of the Social and Behavioural Sciences. Oxford: Elsevier; 2015. pp. 367-373

[38] Birkland TA. Focusing events, mobilization, and agenda setting. Journal of Public Policy. 1998;**18**(1):53-74

[39] African Union Commission. Agenda 2063: The Africa We Want. Addis Ababa, Ethiopia: AUC; 2015. Available from: www.agenda2063.au.int

[40] Kagame P. Report on the Proposed Recommendations for the Institutional Reform of the African Union. Addis Ababa: African Union; 2017. Retrieved from: https://au.int/sites/default/files/pages/32777-file-report-20institutional20reform20of20the20au-2.pdf

[41] Bishoge OK, Zhang L, Mushi WG. The potential renewable energy for sustainable development in Tanzania: A review. Clean Technology. 2018;**1**:70-88

[42] Ozkan G. Global energy security, international politics and renewable energy resources. International Journal of Multidisciplinary Approach and Studies. 2018;**05**(1):39-47

[43] Gordon E. The Politics of Renewable Energy in East Africa. Oxford: The Oxford Institute for Energy Studies; 2018